The End of
the Anthropocene

Ecocritical Theory and Practice

Series Editor: Douglas A. Vakoch, METI

Ecocritical Theory and Practice highlights innovative scholarship at the interface of literary/cultural studies and the environment, seeking to foster an ongoing dialogue between academics and environmental activists.

Recent Titles

The End of
the Anthropocene

Ecocriticism, the Universal
Ecosystem, and the Astropocene

Michael J. Gormley

LEXINGTON BOOKS
Lanham • Boulder • New York • London

Published by Lexington Books
An imprint of The Rowman & Littlefield Publishing Group, Inc.
4501 Forbes Boulevard, Suite 200, Lanham, Maryland 20706
www.rowman.com

6 Tinworth Street, London SE11 5AL, United Kingdom

British Library Cataloguing in Publication Information Available

Library of Congress Cataloging-in-Publication Data

Names: Gormley, Michael J., 1986- author.
Title: The end of the anthropocene : ecocriticism, the universal ecosystem, and the
    astropocene / Michael J. Gormley.
Description: Lanham : Lexington Books, [2021] | Series: Ecocritical theory and practice
    | Includes bibliographical references and index. | Summary: "In The End of the
    Anthropocene, Michael J. Gormley examines late stage anthropocene literature and
    the imagining of the astropocene. Focusing on science fiction literature, Gormley
    frames a changing ecoethic for the end of the anthropocene"— Provided by publisher.
Identifiers: LCCN 2021017907 (print) | LCCN 2021017908 (ebook) |
    ISBN 9781498594059 (cloth) | ISBN 9781498594066 (epub)
Subjects: LCSH: Nature in literature. | Ecocriticism in literature. | Human ecology
    in literature. | Human ecology and the humanities. | Science fiction—History and
    criticism.
Classification: LCC PN48 .G56 2021 (print) | LCC PN48 (ebook) |
    DDC 809/.9336—dc23
LC record available at https://lccn.loc.gov/2021017907
LC ebook record available at https://lccn.loc.gov/2021017908

For Jen and Maddie—
my darlin and my little darlin.

Contents

Acknowledgments

I see no finer place to start these acknowledgements than by thanking the yearling squirrel in my backyard for being hilarious each morning when I am looking out the door instead of at these pages. You bound around in some sort of parkour and playfully harry your peers, sticks, and lawn furniture alike. You are awesome.

I would also acknowledge the authors whose fiction I stitch together in this book. Yes, I plucked what I needed, but I was certain to hold with the underlying themes of the respective works, which are all fantastic. Bits that I strategically leave out or gloss over are likely glaring to those who read these stories, while hopefully remaining unbeknownst to those who have not. Either way, what is here works toward what is not. If you have not yet checked out the novels, prose, and whatever else I made use of, get after them. They are unimpeachable. I would not write about something I do not love and find deeply engaging. So, thanks to all the authors I take up in this work. I am amazed to have found such interconnectivity between your writing. It must be truly human. I half-heartedly apologize to everyone for the times I slipped into another writer's syntax and cadence. I had fun at least.

Thanks Mom and Liss. You spent hours and miles freeing me up to conjure so many words. And thanks Dad. I remember you telling me that I could do whatever I want in this world. I believed you. Now, here this is. Ruth and Bob, thank you so much for your help and interest. So many of these words were written because you gave your time. Thanks to everyone else, too. This book exists because of family.

Gregory W. Sargent, thanks brother. You have been the most critical help I had in the execution of this book. You were the second person I told about every stage, after only Jen. You helped me conjure the structure of this thing just days after I was invited to propose it, and you were the first person to read any of it and then took the deepest dive into its final draft. Scary G, thanks for all that and everything else. Thanks John R. Gallagher. Your interest and support surrounding this book was invaluable. We have shared some very productive times and some very unproductive times. I guess the coolest thing I can say to you here is it's good to have land—obviously. Benjamin Wendorf, thanks bud. I certainly arrived at many of the *Blood Meridian* readings because I had to be able to defend them to you. We have coauthored, team-taught, and share an office. I have also spilled bevvies on your floor (Thanks to Angie, too). Thanks for it all. Opeyemi Odewale, your book recommendations and answers to my sporadic and random microbiology questions were unbelievably helpful in making scientific what I knew about the microbiome from literature and living. Thanks, bud. Special thanks to Cynthia J. Murphy for the enthusiasm, opportunities, and introductions you gave me for this book. It was so funny to bump into you right after I proposed this thing, and I hope this conveys my appreciation for everything since. I am particularly thankful to Robert McHale. Your expertise is a world away from my own, and I was able to confidently move forward by your insight. Thanks to Drew Wisocky. I think the tone I write with is most practiced by my conversations with you. You have been here the longest. Thanks Evan Hatter. You were there when I flipped the switch from law enforcement to literature, maintaining engaged support the whole time. We have walked a long way together.

Thanks to Tom Brown Jr. and the entire Tracker School family. Tom, my first week in camp, you told me to thank Grandfather. I did, and I do. You deserve it too. These ideas are just one of the things that came next.

Thanks to the members of my department, teaching at an institution with no publishing requirements. Instead of that, we have a 5–5 load. We teach more in a semester than some profs do in a year. Arpi Payaslian and Nicole Payen, thank you for encouraging me to announce my contract for this project to the department. Thanks to those in that meeting for the applause, support, and interest. We are at a two-year teaching institution, and it is often hard to think beyond Freshman Comp. Thanks everyone for the earnest interest and encouragement. I would also thank the many folks in the wider college community who have taken interest in this book and given all manner of support. Encouragement from colleagues and students means so much when the work-life balance would likely come easier if I just put my head down and treated this gig as a nine-to-five.

Thank you to anyone who reads *The End of the Anthropocene*. This is my first book, and you opening it means everything. You find me out and about, and the first round is on me. Anyone who bought the book but did not read it, that is awesome too—partly because maybe I make some money off this thing, and partly because I count every sale as an engaged reader agreeing with every word (and refuse to admit any other interpretation of reality). In return, I toiled over the spine so it looks tight on your shelves. A special thank you to anyone who read these acknowledgments. That is a weird kind of cool, and I would love to hear more about you.

Thank you, Jen. Thank you, Maddie. Hopefully, this book helps me return the support you gave to make it happen. This is for us, for conjuring the rest of our lives. Love you, girls. Thanks.

—Michael J. Gormley

Chapter 1

Ecocriticism and the Universal Ecosystem

There is a story about wolves shaping rivers. In this story, we have lived for years without wolves in one of our most celebrated natural, maybe even wild, places. A lifetime ago, we hunted and killed the last wolf pack. Some pack-less remnants may have still wandered for a while, but they are now all gone. We made wolves extinct there. Then, we put them back. (We are at the apex of so many things.) I cannot say how important it was to do this—things can get murky at this scale—but we all know it was wrong that they disappeared and so it was, at least, not wrong for them to return. The wolves did what apex predators do, and herbivores did what they do, and the plants could then do more too. All things living in that place did everything together. And then a river did what it does, but after a while it did it where it used to. The river moved. It now flows over the places that it did before the wolves were gone. And we did what we do: found wisdom and hope in this story.

There are other ways this story is told. In 1995, thirty-one gray wolves are relocated to Yellowstone National Park from western Canada, almost seventy years after the last Yellowstone wolf pack was killed. Roughly a decade later, a study of aspen height in the park reveals that aspen stands along river and stream banks display less browsing of plant material than their upland counterparts, indicating that elk and other ungulate populations are no longer frequenting these riparian aspen areas (Ripple and Beschta 2007). These areas have notable view and escape impediments for elk and, with the return of gray wolves, now constitute "landscapes of fear"—areas in which prey modify their behavior; in this case, reducing time and frequency in the area in response to the presence and newly elevated efficacy of predators in the altered terrain (Beschta et al. 2018). The reintroduction of wolves to Yellowstone began a trophic cascade that redistributed elk populations to more secure locales away from wooded stands along rivers and streams.

This allowed significant growth and recovery in aspen, willow, and cottonwood populations whose root systems stabilize stream and river banks into healthier manifestations and improve their connections to historical floodplains (Beschta and Ripple 2018). Ultimately, the reintroduction of wolves to Yellowstone's large predator guild initiated a top-down trophic cascade that modified the behavior of ungulate populations, primarily the elk, to recover the growth of woody deciduous plants in riparian areas whose root systems improved stream banks and seasonal hydrologic activity. Wolves impact waterways.

The telling of the trophic cascade initiated by the reintroduction of gray wolves to Yellowstone happens in scientific journals, on Yellowstone National Park's website, formally and casually in classrooms, and on news and personal social media pages. Each iteration carries its own rhetorical tropes, most often a sense of wonder that is decidedly positive. There is some disagreement about the reach and potency of this trophic cascade, although much of it either takes issue with the dramatic rhetoric surrounding the shifted location of rivers or the experimental limitations, and thus scientific certitude, of ecological studies of landscapes. Fleming (2019) voices concern that the titular and larger rhetorical frame of Beschta's (2018) findings are problematic, although this seems more related to sensationalist interpretations in popular culture than a problem with the science. Further, Fleming argues that the methodological framework for assertion is problematic in that the claim is based on studying a singular example, thus lacking fundamental scientific replication. Beschta replies that such replication is simply untenable for the environmental sciences (2019). Indeed, how is one to replicate a complex ecosystem in the Earth system, elsewhere and in statistically significant quantities? Still though, wolves are but a piece of the ecosystem.

My own narrative of this research omits parts of the larger ecological interactions in this trophic cascade. I do not mention beaver, bison, and various plant species, as they are the less dramatic notes of the cultural narrative and outside the scope and motifs of this book; the sources cited do take the time. Beschta and Ripple (2016) provide a measured and deeply researched perspective that lucidly describes the complex ecological interactions without overextending or dramatizing the findings. Even growing the narrative to interrelate these species is an oversimplification that precludes the impact of storms, drought, disease, fire, large climatic shifts, and the information from semi-comparable studies on how apex predators contour ecosystems distant from Yellowstone (Peterson 2017, 69–80; Alston et al. 2019)—all of which is murky enough before invoking the vast sociopolitical factors that contort wolf populations and ecosystems. The idea that an apex species can repair dissonant ecologies resonates with exigent human responsibilities in the Anthropocene—more so when the underlying narrative is that a species is innately such a warden. There is a certain mythological tone or allegorical hopefulness contained in human fascination with top-down, single species trophic cascades that rehabilitate an ecosystem.

This book begins with an amalgamated myth and summary of scientific findings of a trophic cascade—that notion that a change in an ecosystem's top predator(s) shifts the population and habits of other species to alter and regain the historical interrelation and structure of the ecological system—to examine the confluence of current social thought in the Anthropocene. While I support continued study of this trophic cascade and am fascinated by its underlying tensions, *The End of the Anthropocene* is not particularly concerned with what nuances ultimately stand as observable biology and ecology. Instead, this trophic cascade is invoked to represent narrative and science in our culture to understand the utility of ecocriticism for humanity in and beyond the Anthropocene. Playing, for the moment, fast and loose with the scale of terraforming, the wolf-induced trophic cascade is a site of "Terraforming as a narrative, a motif, and a concept [that] exemplifies the feedback between sf, science, and wider popular culture" (Pak 2016, 2). This swampy confluence circulates and recirculates totems of human imagination—stories, objective reality, and social interaction—to frame the efforts of this book: to interrogate the end of the Anthropocene and its following epoch by a predictive model of ecocritical literary analysis.

The popularity and importance of wolf reintroduction accesses and represents myriad features of humans in the Anthropocene: we recognize science's ability to observe and describe the world, especially as it produces actionable facts; we have the baseline scientific literacy required to be fascinated by and comprehend a vast ecological mesh; there is an imperative surrounding the ability to impact various environments; we understand the importance of ecological balance and recognize that "wildness" is and begets that balance; this acknowledges our initiating role in environmental stress, destruction, conservation, and reconstitution; as well as representing value judgments relating us to nature and how we must act as those who will continue to impact the planetary system, akin or as a force of nature. We, humans, largely understand our relation to the environment in the Anthropocene.

This might be read as highly Americanized and Western, which it is. Partially, this is because my research is in American Literature, and I have always lived in the United States. Certainly, there are elements of privilege at work as I write in the environmental humanities, so academically exigent in the global north. There is further tension in framing this book with the wolves of Yellowstone because "From the beginning, conservation was tied to racist, sexist, and classist notions of wilderness protection in order to serve urban, bourgeois, white men's desire to construct themselves as rugged frontiersmen [Aguiar and Marten 2011; Collier 2014; Thorpe 2011]. Claims of ownership over 'wild' spaces were used to justify land theft based on the concept of terra nullius—that land was empty and available for the taking" (Curnow and Helferty 2018, 148). Reflecting on these tensions, this book prioritizes a few throughlines: ecocriticism's scientific invocation is just as productive with non-Western sciences as theoretical frameworks for the close readings in pursuit of configuring biotic identity—tracking is arguably the first science, established by first peoples and enabled access to the consistent protein and fat sources that grew the brain (McDougall 2010, 22), and the natural navigation that, through recognition of biotic relation, enabled Polynesian voyagers of the South Pacific Islands to maintain a cohesive cultural identity (Barrie 2014, 264); in and beyond the Anthropocene and the universal ecosystem recognized later in this chapter, differentiation between earthly species will be rather moot; and this book also describes an ecocritical process for onto-logical reconfiguration that subverts disruptions to the formation of ecoethical ontologies.

The End of the Anthropocene interrogates how we read after we acknowl-edge the character of the Anthropocene, and how we read as biotic beings capable of conceiving the end of the Anthropocene and the contours of the epoch beyond. Especially as the Anthropocene and the epoch after concern biotic humans in and beyond the Earth system, the ecoethical ontology we configure in the present describes our actions in the universal ecosystem of which our blue planet is but one part. How we handle Earth's ecosystem is how we will handle others. By this, a primary throughline of the book is establishing a perspective for how our species can approach speaking as nature—that is, maintain the Earth system in scalable agential assertions of self. Literature conjures several beings who just might be capable of speaking as nature, establishing a path for approaching this ideal even while it remains impossible to embody (Buell 2005, 7–8). Our fascination with apex predators and trophic cascades, our fascination with the wolves of Yellowstone should describe a desire for such existence and action.

Most striking about the wolfen trophic cascade of Yellowstone is that it is considered positive precisely because it represents the reversion of the park to a prior, wilder state. Gray wolves were relocated to sustainably regain a moment in ecological history. In their biotic homes, wolves act as wolves which is to act as nature—at least to the degree that any organism who maintains an ecosystem maintains itself. Transplanted wolves replicate a vanished population's ecological impact that describes the natural history of an entire continent. Reinstating gray wolves simultaneously values the action of terraforming the Earth to be more like the Earth and then limits that to those annexed places with distinct borders. Such sociopolitical boundar-ies are just sites of anthropogenic process. When wolves disregard them for biotic reasons they can be legally hunted, even those monitored for study by Yellowstone's rangers and researchers (Blakeslee 2017, 219). At other scales, sociopolitical borders can intensely harm ecological wilds, even those who have found the Anthropocene livable (Miller 2019). Wilderness must be maintained, but only where we want it. Wild places must always be *over there*. Thus, Yellowstone was never "true wilderness, was comfortingly close to civilization, but existed just enough apart to create a boundary. This was what most people wanted: to be *close to* but not *part of*. They didn't want the fearful unknown of a 'pristine wilderness.' They didn't want a soulless artificial life, either" (VanderMeer 2014c, 81). Ultimately, national parks are just urban ecologies, those whose contours are at the behest of social institutions that attempt to make the best of our present for some sort of ecological future.[1]

Human-induced trophic cascades and urban ecologies embracing wildness, I suspect, eventually flounder when they run up against anthropocentric sites that have always sought to make malleable the landscape. A fascination with trophic cascades rejuvenating Yellowstone fails when we ask, what next? Even if the park could become a perfect reiteration of its wild self, I doubt humanity will be reseeding and reintegrating with a wild world. We are in the Anthropocene, a time of disproportionate, if not irreparable, human impact on our biotic home that might just constitute the sixth mass extinction in the history of life on this planet (Ellis 2018, 112; Kolbert 2014). We do not want wilderness to happen here, where we are. Wild spaces should be conserved and visi(ta)ble and over there, anywhere else. We will not get our planet back, not really. The discourse centers on global temperature increase (Ellis 2018, 97), but what about prevention of such a cataclysm revives extinct species and their missing environments and the myriad ecological interrelations all organisms imply? How do we reinstitute such ecologies in the face of over-population and food production? Might we even be too late to prevent Earth from getting any worse before the planet reaches the novel equilibrium of this epoch? The Anthropocene is a new planetary ecology in need of a process for generating a human ecoethic.

The hopefulness inspired by wolf and water recovery in microcosm masks, by our capability to revert landscapes to moments in their biotic history, that we do not really want this effect at the macroscale. Thinking that the trophic cascade from reintroducing gray wolves in Yellowstone is indicative of any desire to terraform our planet to any previous iteration is that same mythologization that scientists seek to combat by taking semantic issue with the titling and telling of those findings. We are not bringing any version of our planet back. Nature will continue to unfold as the product of interrelation between organisms and environment. What matters is that the proliferation of a human-induced trophic cascade occurred in pop culture and science, both. Instead, let us accept such swampiness to access the character of things rather than their minute delineations. This ability is critical because we need to understand that we cannot exit the environment, the universe, the Anthropocene—or the epochs coming—at least not without extinction. Nature is an unimpeachable supersystem not limited to our planet. *The End of the Anthropocene* characterizes the roiling murk of our current time and place. Such an act is the preface to the investigation of humans beyond the Anthropocene and what we will be when we get there.

THE ANTHROPOCENE: SOME BOUNDARY OR A STARTLED SENSE OF DOUBLING

This book argues a biotic human trajectory that extends across geological epochs by offering an access point to the roiling murk surrounding humanity in and after the late-stage Anthropocene. We have a scope of existence inherited from the destructive ecological impact on Earth with a distinct trajectory outward, beyond our planet, and into truly wild environs. We are deciding how best to deliberately manipulate our planetary ecology and live on others. The primary concern of our age is how humans relate to our environment; the underlying theme is the interconnection of all things. *The End of the Anthropocene* progresses ecocritical theory to interrogate the late-stage Anthropocene and witness the epoch after. Recognizing the character of the Anthropocene reveals what sort of act ends the epoch and initiates the next. Conveyed by literature circulating with the construction of ethical and ontological humanity in response to our dissonant environmental relations, this analysis offers an approach for ecoethical interrelation to imminent environs in and after the Anthropocene.

Our approach for examining the end of the Anthropocene must be elegant and simple, and it must be robust enough to function as itself in each location, scale, and discipline without metaphoric slippage.[2] We know that both biotic existence and the various disciplines of knowledge are "a vast sprawling mesh of interconnection without a definite center or edge" (Morton 2010, 8). This notion is equally descriptive of philosophy, ecology, and the universe. Wonderfully, this ecological thought makes no real distinction between them either. Infinite mesh, however, is difficult to imagine, to conceptualize or realize, and without careful footing leads to dangerous approximations or disengagement (Clark 2018, 18). *The End of the Anthropocene* entrenches ecocriticism in the transitional space between the environmental humanities and the environmental sciences, drawing in disciplines like evolutionary biology and physics while never straying too far from literature to approach an ecoethic initiated by biotic identity.

Biotic identity—the self as described by the physical interrelation with one's ecosystem and the other organisms within—is currently a novel entangling of Holocene adaptations in the ecosystems of the Anthropocene. The relative rapidity with which the average Earth temperature has increased signals anthropogenic impact at the global scale (Ellis 2018, 66–68). The Anthropocene creates novel ecological interrelations between organisms evolved for the ecosystems of the prior geological age. In the Anthropocene, the bidirectional biotic maintenance of organisms and their environs is restructured. Whatever event ultimately demarcates the Anthropocene, this epoch is defined by human impact on Earthly environs as a force of nature in the total planetary system (Ellis 2018, 14). This anthroposphere distinguishes the current planetary ecology from the prior and has already reissued the biotic identity of all species.

There may be revenant organisms and places unimpacted by the anthroposphere, but the containing macroscale Earth system has been definitively restructured by anthropogenic impact. Just as the reintroduction of wolves did not create a primordial Yellowstone that seeps outward to make wild the world, organisms maintaining their pre-Anthropocene biotic actions are now included in the environs of the Anthropocene. Such organisms might reinforce the Anthropocene's environs or their actions just do not scale to global impact. Environmentally conscious individuals are recognizing their similar biotic identity. Acting by an environmental conscience does not necessarily create systemic change.[3] In the Anthropocene, organisms are doubled, metonymically referencing the prior ecosystem from which they evolved and the current anthropogenic ecologies in which they are entangled. Turning to Jeff VanderMeer's *The Southern Reach Trilogy* provides a theoretical framework for interrogating such characteristic human doubling as it details a species out of place in its own ecosystem and simultaneously describes the mysterious Area X and our Anthropocene.

Ever concerned with doubling—as a sensation and actual, literal replication—VanderMeer's trilogy uses this motif as a process of Area X, terrifying and destabilizing characters and readers interacting with the strange and wild place. In the novels, Area X forms a creature of its first human contact who writes an apocalyptic sermon in English words on walls in sporing flora and microorganisms; dolphins with human eyes swim upriver while meeting the gaze of the biologist; and some former human sheds face and skin like an exoskeleton, "detritus from a kind of molting: a long trail of skin-like debris, husks, and sloughings" (VanderMeer 2014b, 322). Helpfully, Garland's 2018 film interpretation of the trilogy's first book describes Area X as "a prism, but it refracts everything": GPS, radio waves, cell structure, even DNA. In the film, a bear with a decaying or growing face consumes a character's throat and mimics her screams to hunt her companions in a living room while just outside plants have taken human shape because their cells are the same cells that give us ours. Constituted by biotic process, that cooperative exchange of shape by organism and environment, each of these creatures is metonymically Area X and display the characteristic doubling at the core of this environment's interactions. Area X puts in flux all physical organisms it encounters, and in doing so, looks very much like the Anthropocene.[4] Ultimately, the ecohorror of Area X is the biotic dissonance of human organisms interacting with a new earthly ecology that is not for them and within which they have no authority.

In *Authority*, the second of *The Southern Reach Trilogy*, while making his way into the pristine wilderness consumingly encroaching upon Earth's environs, Control experiences "a startled sense of doubling, as if he were somehow traveling up the coast of an alternative Area X. The sense that he had passed beyond some boundary" (VanderMeer 2014c, 322). Approaching Area X, Control experiences the otherworldly wilderness biotically, bodily relating to the new environment rather than logically charting the invisible border. Suddenly incorporated in the ecology of Area X, Control is doubled. He is an organism produced by exchanges with the Earthly environment he no longer inhabits, and this lack of authority characterizes his often-impotent actions for the remainder of the series. This subtle scene is *The Southern Reach Trilogy* at its least fantastical. Control becomes a creature of Area X and is now subject to an ecosystem within which he has no authority or innate, historic relation. This notion of doubling, read to inform human experience in the current epoch, demonstrates the core axiom of the Anthropocene: humans impact, without controlling, ecologies.

Still holding in mind the wolves of Yellowstone, I would recirculate them with Control's otherworldly sensation of doubling in Area X as a framework for characterizing the Anthropocene. I recall a striking video of a couple driving through a snowy Yellowstone and encountering a pack of wolves.[5] The video opens on a black wolf trotting down the road toward the vehicle and passing by the driver's door with just two slight glances. In the small, wide-angle mirror, we catch a glimpse of another wolf waiting close behind the vehicle for its companion. Together, these two casually move off to rejoin the pack, walking in the tracks of other animals and vehicles that have come through since the snowfall. The camera zooms in on the larger mirror and shows that nearly the entire pack is waiting in the road with a few members remaining uphill, looking out from the tree line. Another wolf stands both in and out of the vehicle's snowy tracks and howls while gazing toward the returning pair, toward the vehicle and camera. The wolves occupy the plowed and newly snow-covered road as though it were another, any other, traversable feature of the mountainous terrain. They act with the aloof multidirectional awareness of a pack of apex predators in an area of high visibility, unaltered by the supposed human authority and control asserted by the road or vehicle. The couple speaks in whispers as more wolves add their howls.

This image of Yellowstone's most iconic creatures displays the doubling inherent in human interaction with anthropogenic environs. Those in the video whisper, readable as awe, unease, and a desire not to disturb the scene. They are doubled just as Control is doubled, responding to the suddenness with which the area has become otherworldly, wild. The scene houses myriad doublings beyond wolf and human—which should not be taken as simply nonhuman and human. Rather, this doubling reveals those living well with the park. These gray wolves are descendants of the imported Canadian pack yet double as the very wolves hunted and exterminated. Yellowstone's other-worldly tone comes from the double vision of witnessing a pre-Yellowstone, even pre-United States, wilderness in the park on a road that has itself been apocalyptically doubled as the wolves act as wolves in, around, and despite the human structure. Contained in this video are the narratives that began this book, of humans impacting, destroying, and conserving ecosystems. Watching, we feel as humans within and without wilds. (The deconstructive potential of this scene is a pleasant sort of murk, staggering though it may be.) If we pay close attention, we see in the park's construction the same reason we read VanderMeer's work. Humans attempt to manufacture ecological control as we experience an ecogothic and very real uncontrol, a doubling in the environment.

These doubles, by existing biotically, resist becoming deconstructed binaries and instead reveal an ecocritical metonymy. Categorical binary oppositions dissipate as individual humans are doubled by a single process—unprecedented biotic relation to an environment—and reappear as something in a broad array of possibilities. VanderMeer's Control, the biologist, the biologist's double, every other character, or anyone in Yellowstone sit as members of a single species variously and inequitably equipped to survive. An organism and all other organisms relating to an environment by speciation is far too complex to be modeled as binaries such as organism/environment, organism pre/post environmental shift, or even successful/failed integrations. The doubles are produced by exchanges with their environment, which is another chaotic myriad of interactions constituting an ecosystem. Even then, that biotic self is only conceptually altered. Counterintuitively, the doubled self is organically unchanged as evolutionary adaptations that configure biotic identity emerge from within a species, not established organisms. Rather, the organism is recategorized in terms of potential to survive in the environment. Delivering on that potential constitutes the ecology of a place. Nothing is deconstructed without anthropocentric narrative, and ecocriticism casts that off.

Additionally, ecocriticism eschews the luxury of finding deconstruction literally in death as a dismantling or disintegration, and instead reinforces metonymy, that conveyance of accessing the whole by its part. In the landmark "Greening of the Humanities," Parini describes the axiom with which we read ecocritically: "From a literary aspect, [ecocriticism] marks a re-engagement with realism, with the actual universe of rocks, trees and rivers that lies behind the wilderness of signs" (1995). Adding death to this list of abiotic factors converts it to complete necrotic decay (later read as entropy). Ecocritically stripped of connotation, death is merely another perfectly acceptable integration with the environment, even at the scale of extinction.

Deconstruction arrives at the same conclusion by finding no difference between integration and disintegration, but so does grade-school biology and *The Lion King*. The decay of an organism is natural and inevitable, always already an organic process constituting nature as itself, the unimpeachable supersystem. (What could be the binary opposition of nature? Of anything that contributes to ecology?) Metonymy is a literary process unimpeded by ecocriticism's unconcern with metaphoric signals. All forms of doubling in an environment, even those organisms who die, produce a metonymic relation that traces an organism to some set of environmental factors. Humans are simultaneously a representation of imbalance in the environment and an extension of perfectly organic processes by which a planet settles after any event that could constitute epochal naming. We metonymically trace our species back through various anthropocentric hyperobjects—those things we create or manipulate which are "massively distributed in time and space relative to humans" (Morton 2013, 1) and "do not rot in our lifetime" (Morton 2010, 130), such as radioactive materials and plastics—and find the anthroposphere. Ecocritical metonymy reveals the anthropocentric disconnect between our ability to impact the planet and inability to control the resulting ecology of the Anthropocene.

The Southern Reach Trilogy images an Earth ecology that is become something else and provides ecocriticism a near-allegorical framework for assessing the character of the Anthropocene through the nonhuman and human animals within the epoch.[6] Area X carries an unsettling ecohorror in its verdant desolation. This eerie replicating of the Earthly location causes for the biologist "the unsettling thought that the natural world around me had become a kind of camouflage" (VanderMeer 2014b, 98)—Control's same doubling experience filtered through her expertise. She is inextricably within an ecosystem that appears as the place it no longer is, thus stripping her of intellectual authority derived from any prior, observation-based scientific praxis. Instead, the biologist's relative success in dealing with Area X comes from her ability to integrate into landscapes and the literal inhaling of its microorganisms and processes. The earthly environment is supplanted and mimicked, doubled and converted by Area X.

The place is still organic and earthly enough that the expedition is "assured it was safe to live off the land if necessary" (VanderMeer 2014b, 4). Area X is biotically habitable in extremis—deriving anxiety from a human becoming inextricable from a wild place, even though it might lead to living well with the environment—but is sufficiently altered so that human organisms experience on all sensory levels the place's otherworldly *wilder*ness. This lack of control that was supposedly granted by being a nonanimal human—which is, of course, not a real category—is precisely the terror of the trilogy. The ecosystem proceeds outward by an uncertain and indefatigable process. Area X is encroaching, undeterrable (VanderMeer 2014c, 303). Earth is becoming Area X as Area X becomes the Earth. Creatures are doubled by existing, biotically interacting, with and in a new ecology by bodily adaptations for the planet's former ecosystems. And there is the allegory: Area X is all anthropogenic sites. Area X is the Anthropocene at all scales—bordered wilderness preserves, this planet and others, the universe. Area X, Control, and the biologist uniformly map to the experience of humans relating to the environment in the Anthropocene, and after.

In the Anthropocene, human impact constitutes a planetary ecology that no longer secures our species' survival. Or, as the biologist's double puts it: "The world went on, even as it fell apart, changed irrevocably, became something strange and different" (VanderMeer 2014a, 328). We now experience the reciprocal nature of environmental impact—we impact environments and they impact us—and find that our actions have all but guaranteed a nonhuman future for the planet. Erle C. Ellis, in his brief introduction to the Anthropocene, details that our global system—the interrelation of the bio-, litho-, hydro-, and atmo- spheres—now includes the anthroposphere, stating that "Nature untouched by humans had now disappeared through the global reach of a human-altered climate" (Ellis 2018, 14). Or, from *The Southern Reach Trilogy*: "You've never walked through an ecosystem that wasn't compromised or dysfunctional, have you? You may think you have, but you haven't. So you might mistake what's right for what's wrong anyway" (VanderMeer 2014a, 32). The same tendrils drawing together Area X and Yellowstone characterize the dissonant human-environment relations in the current epoch. The anthroposphere is a great force of nature disrupting, altering the pre-anthropogenic functions of the planet. The dissonant doubling of humans in environs like Area X and Yellowstone is a product of our "disproportionate impact on the planet" (Schaberg 2020, 152). The landscape changed, and we now live in a Darwinian ecogothic plot where the physical adaptations we have were designed by an environment that no longer exists. In the Anthropocene, we experience the ecological duality of *impact*. The impacted planet impacts its organisms, displaying the vast difference between our ability to impact nature and control it.

SCALING THE ANTHROPOCENE

Deriving an ecocritical framework from VanderMeer's doublings is especially productive for considering the reciprocating consequence of humanity impacting and being impacted by the Anthropocene's ecosystems, particularly in response to Clark's jarring and inspiring call to reframe ecocriticism. Clark describes a "conflict of scales" (2018, 23)—discord from ecological solutions at one scale becoming harmful or impotent at another—which Clark shows also occurs when ecocritically reading literature in and for the Anthropocene (2018, 10). Reissuing ecocriticism with *The Southern Reach Trilogy* resolves the scalar dissonance, first and most dramatically by not resolving the planet's ecological shift. Scaling Area X to the Anthropocene states that we cannot bring the Holocene back. The anthroposphere marks a barrier between our ecological present and the pre-Anthropocene planetary ecology to which we do not return. In the Anthropocene, we are horrifyingly doubled as we face a planetary-scale nature that we coauthor but maintain no authority over. The scalar discord Clark describes is rooted in the perception that controlling one's actions controls the anthroposphere (2018, 140–141). The organism functioning singularly at the Darwinian-Malthusian scale of individual interaction mistakes those actions for systemic impact on the anthroposphere (Hagman et al. 2019; Mann 2021, 60–62). Alternatively, *The Southern Reach Trilogy* dissolves human ability to control the environment, leaving only integration, either by living well or through postmortem decay. The anthroposphere is the product of a species of biotic constructs responding to their immediate environs, not an individual responding to global ecology.

Instead, *The Southern Reach Trilogy* collapses the ability of individual human action to impact ecology in favor of a vision of biotic integration with the immediate landscape. We are in the Anthropocene and have not yet committed as a species, as a global culture, to positive and perpetuating integrated living within our environs (Mann 2021, 1). We mistake individual action as scaling to impact the contours of the anthroposphere. But, the anthroposphere altered the baseline of the planet. Any further action happens within the Anthropocene, not within or toward achieving some pre-Anthropocene ecosystem. *The Southern Reach Trilogy* instructs that human future on Earth is at stake, not the planet's ability to maintain some version of the biosphere, of life. The anthroposphere does not dissipate and we do not want it to, not really. Even efforts to artificially lower global temperature maintains the Anthropocene, albeit with a sustainability akin to pre-Anthropocene ecosystems. The anthroposphere will persist, even with immediate human extinction. The ethos shaping the anthroposphere must be thoroughly concerned with perpetuation by integration.

The Southern Reach Trilogy catalogues the success (and character-foil alternatives) of the biologist's literal double integrating with an untransmutable ecology "to become so attuned to my environment that after a time no animal, natural or unnatural, shied away from my presence . . . I grew attuned to [the plants'] messages as well" (VanderMeer 2014a, 177–178). She integrates with the wild so thoroughly that she even communes with plant life, becoming a wholesome feature in the immediate ecosystem of Area X. Demystifying communicative interactions with plants, Haraway explains that "Plants are consummate communicators in a vast terran array of modalities, making and exchanging meanings among and between an astonishing galaxy of associates, across the taxa of living beings. Plants, along with bacteria and fungi, are also animals' lifelines to communication with the abiotic world, from sun to gas to rock" (Haraway 2015, 122). Changing her actions to live well with the immediate environment because she cannot undo the new Earthly ecosystem dissolves the terror at the doubling and creates her as a positive metonymic expression of her environment, or a metonymic expression of positive ecological engagement.[7] Becoming receptive to the metonymic pathways by which organic beings imply their biotic relations and the abiotic contours of the environment inculcates one in the myriad interrelations sustaining environments at all scales. The plants and animals imply each other, their necessary interactive maintenance, and the abiotic features of the environs. Even extending to the continental, atmospheric, or extraplanetary scales is authenticated by the individual organism's metonymic entanglings, its biotic identity. I reckon these as proper pathways for ecocritics to travel as they pursue the larger goals of the environmental humanities, namely, some unified ecoethic governing the anthroposphere's impact.[8]

Issuing *The Southern Reach Trilogy* as an answer to Clark's call to reframe ecocriticism allows a reembracing of Morton's ecological thought (2010). Ecocriticism finds in the doublings of VanderMeer's trilogy a framework for the disconnect of human ability to impact the planet. Recognizing this allows a proper tracing from human to the Anthropocene, to situate the discord as a manifestation of Clark's scalar problem. Humans experience the simultaneity of our actions having already scaled to the planet while our individual actions (and maybe any further action) do not. At each scale, *The Southern Reach Trilogy* expresses the same conflict. Humans find themselves adapted to an environment other than the one they are in. There is terror at the terroir—the ecological interconnectedness of all things. In *Authority*, Whitby defines *terroir* to Control: literally translated as "a sense of place" but referencing an organism as a unique entity whose genetic identity is deeply entrenched in "the specific character of a place—the geography, geology, and climate" (VanderMeer 2014c, 131). The concept originates in deeply ecological winemaking perspectives but is scaled to describe Area X.

When applied to ecocriticism, we find that we already have this idea—Morton's ecological thought—and we have been applying it for a while. Clark is particularly skeptical of this notion, wondering "What if the kind of transformed imagination celebrated in this sort of cultural programme, this awareness of interconnection, could not be assumed to be an effective agent of change—in other words, how *far does a change in knowledge and imagination entail a change in environmentally destructive modes of life?*" (Clark 2018, 18). If, however, we begin in *The Southern Reach Trilogy*, we gain sturdier footing when witnessing the interconnectedness of the biotic and abiotic all. While maintaining the lack of centrality inherent in the concept, we still begin somewhere allowing the ecological thought to resolve any remaining anthropocentric baggage acquired by metonymically tracing to ecosystems from human. Further, that beginning is self-reflective and resistant to anthropocentric restriction. This perspective is necessary, for the true scope of this book is to prepare humanity for its fast-approaching integration with extraplanetary ecosystems.

Ultimately, Area X is alien. The otherworldly environment consuming, replicating, supplanting, and already becoming our world came from an extraplanetary elsewhere. In some way, that other world is connected to Earth through a literal "worm" hole, terrifying precisely because it does not look like classic SF wormholes, that oddity of optics that feels of void more than nature. Instead, VanderMeer has doubled that image into an organic, esophageal, terraforming entity—which is just a worm that has been scaled up to double Earth into material that nutrifies the environment in whichever incomprehensible way Area X uses resources. There is a disturbance when wormholes acquire the trappings of organic systems as opposed to the quiet, striking beauty of the no-less-natural extraplanetary universe. As these ideas scale, they become characteristic of human interactions and help ecocriticism to collate temporally and offer proper thought to future biotic interactions, specifically as we exist on the cusp of colonizing the moon and Mars (NASA 2020c).

Reissuing ecocriticism from *The Southern Reach Trilogy* in response to Clark is critical to the tremendous spatiotemporal scope this book takes up. Humanity is already a species that does not, in total, live on this planet. Soon the same clock recording our species' extraplanetary history on the International Space Station will start for the moon and for Mars. What it means to be human will be altered as the species begins existing places other than Earth. Humanity will witness a reiteration of the Anthropocene, but with foresight. Elsewhere in the universe, we will again find that a planet is not for us and will attempt to survive that very environment. The Anthropocene is a strange and startling border. The minutiae of the epoch are "largely incalculable" (Clark 2018, 1) but the structure and character—the core, the bones, the processes—are not. The boundary that begins the epoch is the anthropo-sphere, and the sensation is being able to impact the planet without control over the output. This book finds the end of the Anthropocene in imagined realities circumscribed by the objective realities of ecocriticism to establish the contours of humanity as we become a species unbound from the blue planet on which we began.

THE (A)BIOTIC NOTION OF HOME

The notion of our biotic home—that blue planet we evolved to suit—is now common enough. All species exist in bidirectional impact with the other organisms and terrain in their respective regions, and the network of these ecosystems constitutes the Earth system. The Anthropocene is but a new iteration in which humanity finds itself dissonantly interrelated with its environment. While this notion is most ardently circulated in relation to the climate crisis, science fiction has also been doing this sort of thing for a while now, often putting humans on alien planets or in survival situations in and outside of spacecraft. SF characters frequently find themselves hunted by alien apex predators or stumble across some spore or chemical compound or excessively wet egg that messes with human biology. SF humans regularly run low on oxygen or whatever other need-it-to-live resource. Occasionally, artificial gravity is compromised. An SF series might not be doing its job if it does not take at least one episode to invoke such tropes.[9] Such dissonant interactions between humans and environs are the "substratum" of our culture (Schaberg 2020, 152).

These tropes are hardly novel representations of how science fiction recirculates with the present. Even the mythos of Apollo 13 is a fairly stock iteration of the circulation between real and SF life-support failures. The xkcd comic "The Martian" riffs on the idea that Andy Weir's, and thus Ridley Scott's, *The Martian* is for people who watched the scene in *Apollo 13* "where the guy says 'We have to figure out how to connect *this* thing to *this* thing using *this* table full of parts or the astronauts will all die'" and then wished "the whole movie had just been more of that scene" (xkcd). The punchline is the characters wondering how it became a Matt Damon blockbuster. *The Martian* describes a human in the immediate future trying to survive on a planet, which also describes the immediate present. The film offers that, with our technology, we can simulate Earthly environs to survive. In such narratives are a biotic home to which the astronauts can return and live extricated from spacecraft and suits that replicate minimum survivable earthly environs—abiotic factors like atmospheric pressure, breathable air, temperature. Common as they are, these ideas describe the very fabric of the late-stage Anthropocene: controlling human (in)compatibility with the environment. The anthropogenic impact of the Anthropocene, however, has challenged the stability of planetary features such as climate and air content. All we are doing now is finding that we can use the SF imagination to alter our planet to secure the abiotic features our biotic selves require. (Geoengineering is very SF.)

Our biotic identity—that identity formed by relation to other living organisms—always includes an abiotic context of nonliving environmental factors such as weather and climate, landscape features, water and other molecular and mineral resources. Ecosystems are constituted by the reciprocating impact of biotic and abiotic factors.[10] Darwinian and Derridean, biotic identity reveals environment. The characteristics of the external environment are directly referenced in those of the organisms within; even the failure of a species entire describes the ecosystem itself. Deliberate ecological impact in the late-stage Anthropocene focuses on manipulating the abiotic factors of the planet to suit our biotic features, theoretically also suiting whatever other biotic relations we need. Our species, like all species, survives or perishes based on interrelation with biotic and abiotic contours of our home planet. Reacting to SF tropes manifested by our home planet ecology—the one place climate and air content should not be a concern—the late-stage Anthropocene is when we control our impact on our biotic home. At this end boundary, doubled on our own planet, humans deliberately shape the contours of the anthroposphere to make Earth bodily habitable. In and after the late-stage Anthropocene, humans react to the experience of reversion to organic bodies evolved to survive within the relatively narrow limits of a distant Earth ecology.

Never has human physiology been so dramatic in its metonymic expression of our biotic home. Fast approaching the late-stage Anthropocene, we are simultaneously engaging in planetary-scale destructive environmental practices while preparing crewed missions to explore and colonize the moon and Mars (NASA 2020c). In all locations, current and forthcoming, human bodies are not at home. Since November 2, 2000, we have maintained a permanent extraplanetary—literally extraterrestrial—human presence on the International Space Station. We are a species that no longer lives exclusively on our biotic home. Soon, that clock will start for the moon and for Mars. Due to the Anthropocene's disruption of the previous function of our biotic home, our species is now alien everywhere it is and everywhere it is headed.

Our current extraplanetary excursions are largely concerned with the abiotic factors needed by humans. On the ISS we import and manufacture breathable air and water, maintain necessary climate and atmospheric pressure (NASA 2017b), while we just mitigate the impact of microgravity on the human body (Garrett-Bakelman et al. 2018; NASA 2021). We are preparing to give the generation of children huddled around TVs for Armstrong's "giant leap for mankind," my parents among them, the same moment on Mars—after, of course, returning to the moon to stay. Even if these colonies exist in places that do not have a biosphere of their own, they are not our biotic home and maintaining even the minimum survival resources will be a consuming need—an exigency eerily applicable to our earthly environmental interactions. We are preparing to permanently reside on other near-Earth objects and culture needs an access point to describe humanness as it shifts to something novel and unprecedentedly physiological. I find it simplest to just stare at the dirt and go from there.

For around ten years now, I have been studying tracking—both as a discipline in its own right and as a critical framework in literature and science. I was taught to track on the week-old trail of a young deer meandering over debris and by following the path of a fox opportunity hunting across moss. I tracked a deer in the dark using my fingertips and watched the world-renowned tracker who founded the wilderness survival school I attend scrape away layers of packed earth to reveal rodent and deer tracks a decade old. Once, I even got to participate in some forensic tracking as some backwoods locals set spike traps to make the area's weekend partiers feel unwelcome.

I am no master tracker, but I have more dirt time than most people, and I have been lucky enough to stare at the ground next to experts instructing how to see what once stood there and what it was doing. Tracking is as artistic as it is scientific with more than one field guide title offering tracking as art and science, both.[11] As a literary trope, tracking ranges from investigative plot device to a near-magical witnessing of the being that stood in the landscape. (There is a disproportionate lack of criticism to the frequent inclusion of tracks in narratives.) While depth and representation of a track's nuance varies from text to text, the information conveyed is always the same: the track is unique to the individual making it and leads only to that organism. Tracks prove the absolute physical actions and existence of a being. Mythical as reading the dirt may feel, tracking is very likely the first scientific effort our species ever performed (McDougall 2010; Liebenberg 2013). *The End of the Anthropocene*, with far more nuance, performs ecocritical close readings of tracks as expressions of biotic identity, the bodily organism entrenched in an ecosystem.

Tracking prioritizes understanding organisms in the larger environmental context, thus describing our role, place, and needs in the Anthropocene. Additionally, tracking functions identically in all disciplines, always describing a body in a place. In ecocriticism or biology, tracking describes the relation between the individual and its environment.[12] Tracking even works on other planets. We left many things on the moon, but with no atmosphere or weather the astronauts' tracks are still visible decades after they were made.

Lunar geologist Noah Petro explains that "We can retrace the astronauts' steps with greater clarity to see where they took lunar samples" (NASA 2017c). I reckon the granular contours show some crumbling, but I would be happy to make the trip and check, if any of you out there can swing that. Other indicators of lunar humans have likely not retained such identifying characteristics in the extreme temperatures and unmitigated solar radiation— the constant abiotic climate of the moon (Spudis 2011). Those tracks are the people who stood there in a way that a rover or flag is not. Tracking is a universal access point to the concerns of our epoch and each discipline weighing in. Tracking is an art and science that grants access to ecosystems. As we spread into the solar system, we quickly become organic bodies in extreme and unprecedented environs. The ontological and political self does not register when you are unsure where your next breath is coming from. Tracks always relay physiology as it relates to environment, which precisely drives human existence in alien environs. By tracking, on this planet and others, *The End of the Anthropocene*'s next chapters establish an ecocritical framework that functions identically in the sciences and literature to interrogate humanity in the universal ecosystem.

ECOCRITICAL PROCESS

The End of the Anthropocene is foremost a work of ecocriticism, thus it always circulates with the Anthropocene. Not yet an official geological epoch, the Anthropocene is a cultural notion born in the sciences and reinforced in the humanities (Mann et al. 1998; Crutzen 2000; Schaberg 2020).[13] The above doubling at its heart, the epoch describes the biotic disconnect that is sorted by evolutionary process—broadly, modification by natural selection or extinction. Engaging large swaths of time, this book requires a robust methodology, able to function pervasively and predictably to reveal the character and processes of biotic identity in and beyond the Anthropocene by invoking usually disparate fields. Ecocriticism is already interdisciplinary, its examination of human relation to nature is an undercurrent in human existence, be it the history and sociology of civilization, literary reflections of place, or biological categorization and recognition of physical interrelation. As literary theory, ecocriticism already maneuvers and conceptualizes human culture in relation to nature and is innately suited to investigate the Anthropocene at its end.

Ecocriticism reveals the underlying character of the Anthropocene— recognizing the biotic doubling of ecological impact and responding with ecological control. The Anthropocene is the effect of human impact rippling along biotic lines. Humans are evolved by the ecological contours of their environment and then alter the environment to, seemingly, better suit the species. This altered environment shifts which other organisms are best suited to survive. Eventually, human impact is global and there is no unaltered environment on the planet (Ellis 2018, 17). Ecocriticism circulates the history of this work and the efforts in response to a planetary environment that no longer maintains the species. With the Anthropocene characterized, a combination of trajectory and ecocritical thought surmises the sort of event so fundamentally differentiated from the current epoch that it cleaves the Anthropocene, beginning the next geological age.

Axioms and Waves

The End of the Anthropocene relies heavily on two ecocritical axioms: that one cannot speak as nature and that ecocriticism is a reengagement of real nature. Of the first axiom, that we might speak a word for but not as nature, Lawrence Buell's iconic *The Future of Environmental Criticism* concludes that "Although there is something noble about human attempts to speak ecocentrically against human dominationism, unless one proceeds very cautiously there soon becomes something quixotic and presumptuous about it" (Buell 2005, 8). For this, Buell succinctly invokes Thomas Nagle, Aldo Leopold, and Henry David Thoreau while also noting how the notion differentiates ecocriticism from gender and critical race theories—the naturalist does not speak as nature where one can speak as a woman, as a transgendered person, as a person of color (Buell 2005, 7–8).

Ecocriticism regards the paradoxical ability to manifest the anthroposphere or caretake the planet without the ability to be as nature. The next chapter engages this more intensely, ecocritically imagining those who might speak as nature (or think or act) and considers characters who embody speaking as nature or find themselves with an opportunity to do so. For that is the point of all ideologies, to hold us on a path to perfection despite knowing the impossibility of arrival. Why else would the word "utopia" exist and be continually reimagined? The point is not to get there, but that halfway from here to utopic is always better. Strive to speak as nature each day and forgive yourself each night when you have failed. Investigating speaking as nature positively contours the shift from unintentional ecological impact to controlling the anthroposphere. Sufficient for now, ecocriticism at the end of the Anthropocene details the doubling characteristic of the current epoch.

Of the second axiom, Jay Parini, in the seminal "The Greening of the Humanities," reflects on the first conference of the equally pivotal Association for the Study of Literature and the Environment (ASLE). The article offers that ecocriticism "marks a re-engagement with realism, with the actual universe of rocks, trees and rivers that lies behind the wilderness of signs" (Parini 1995). By this, ecocriticism is a splendid sort of formalism. The text itself displays literal, instead of metaphorical nature. Certainly, the lightning bolt that strikes the tree in *Frankenstein* symbolizes the illuminating power over the natural world brought by science, such that it could tear it down. Ecocriticism instead pursues the ecological role of lightning and how fire progresses forests. Importantly, the lightning and fire become the likely biotic history inspiring Promethean myth—humans probably gathered and maintained fire before making it. The notion wonderfully arrives at the core concepts of *Frankenstein* without the symbol. Both non-mutually exclusive readings affirm the other. Beyond the formalist value of understanding the text itself, the theory elevates setting and character to investigate the ethics of human interactions with nature. In and beyond the Anthropocene, characterizing the relationship between character and setting is a practical technology for establishing an ecoethic for biotic human relations.

I consider ecocriticism to be the study of literature and the arts within the environmental humanities, a larger category emerging from ecocriticism's increasingly interdisciplinary scope—just as feminist theory conjured gender studies. At the macroscale, environmental humanities and environmental sciences coalesce into environmental studies. In this interdisciplinary schema, disciplines like ecocriticism and biology are free to wander and be wandered into. At the microscale, ecocriticism retains its two waves. First-wave ecocriticism becomes most associated with examining personal relations to nature while second-wave ecocriticism is an analysis characterized by invocation of scientific frameworks for nature. Historically, the waves are far more detailed and both approaches are currently utilized, often overlapping and coalescing. The two waves are sometimes indistinguishable, which is perhaps fine.[14] Although easily parsed by identifying primary texts and objectives, those sites of overlap often define spaces of environmental justice and ethics as they read cultural texts. These overlaps tend to invoke literature but need not do so necessarily. Contrary to reductionism or oversimplification, these interactions are sites at which the environmental humanities become more robust and resilient by tracing connections to additional disciplines. Ecocriticism will still find moments where it is difficult to parse first- and second-wave ecocriticism, but this evokes ecocriticism as a lens through which we relate to narrative and art.

Distinguishing between first- and second-wave ecocriticism in this way aligns the above axioms with the waves. Chapter 3 uses this framework to imbue the two waves of ecocriticism with reciprocating movement to analyze certain hard science fiction texts—those characterized by their dedication to accurate scientific depictions—as personal narratives at the end of the Anthropocene. The ideal of speaking as nature becomes imbued with scientific objectivity as second-wave ecocriticism parses which works of fiction can replicate nature narratives of the future. This move allows examination of the geological epoch and human paradigm after the Anthropocene. By this framework, ecocriticism becomes a predictive theory. These perspectives are necessary to perceive the end of the Anthropocene and characterize the following epoch.

As the temporal landscape of *The End of the Anthropocene* stretches so far into the future, I would see this analysis first rooted in historic and totemic ecocritical approaches. Taking cues from Lawrence Buell's history of ecocriticism (2005), himself a monolith in the framework, I note the proto-ecocritical trends in Leo Marx's *The Machine in the Garden* (1964) and Raymond Williams's *The Country and the City* (1973). From there, I invoke Timothy Morton's *The Ecological Thought* (2010) and Donna Haraway's *Staying with the Trouble* (2016). Temporally balanced by this iconic ecocritical suite of texts, this book approaches the end of the Anthropocene.

Marx and Williams

To reinforce the futuristic extension of this book, the theoretical framework begins where a prototypical ecocriticism coalesces. William Rueckert coins "ecocriticism" in the late 1970s as an experiment with critical theory that assesses the biosphere and its relation to literature (Rueckert 1996).[15] I note the experimental manifestation here as a motif embraced and perpetuated in *The End of the Anthropocene*. Chapter 3 develops a new internal motion for ecocritical analysis to envision the next epoch traced for the rest of the book through invoking physics as ecocritical theory—specifically with descriptions of nature with entropy. In a passing rumination, Rueckert also invokes entropy as a critical lens, positing its utility for Roethke's poetry. As with all literary enterprises, there are prototypical precursors to the minting of ecocriticism.

Buell traces ecocritical thought to first manifest in the work of Leo Marx (1964) and Raymond Williams (1973) as they provide a "history of attitudes toward nature vs. (for Williams) urbanism and (for Marx) industrial technology" (Buell 2005, 14). Regrettably akin to much of literary theory, ecocriticism is wrapped in the history of Western expansion and thought. Resisting this, chapter 2 establishes tracking as an ecocritical framework—a non-Western, natural knowledge base initiated by first peoples and likely the first scientific thought (McDougall 2010; Liebenberg 2013). Chapter 3 similarly reads biotic identity through tracking and invokes the lens of natural navigation, an "art and science" practiced by the indigenous people of the Pacific Islands (Polynesian Voyaging Society; Barrie 2014). *The End of the Anthropocene* traces the biotic identity recognized by tracking and natural navigation into novels set in the natural locations of human futures, continuing the examination of human existence "between settlement and frontier or wilderness" (Buell 2005, 14) that draws the attention of Marx and Williams.

Rife with images of the mediated nature between civilization and the wilds, chapters 3 and 4 examine science fiction tropes of survival and terraforming on alien planets. Although agriculture on Mars and terraforming a post-cataclysm Earth are far grander, Marx and Williams also describe the mediation of wild nature with technology. Marx prefigures the cultivation of Martian crops I discuss in chapter 3 when he characterizes a landscape's abundance as "less closely associated with the landscape than with science and technology" (Marx 1964, 40). My argument extends this construction of the landscape to its totalizing maximum where terraforming a planet resonates with Williams's notion that "The genius of the place was the making of a place" (Williams 1975, 124). There is something innately human in an ecological interaction wherein the species seeks to control the environment. Marx and Williams, science and science fiction are all circulating in the mediated space between civilization and nature that informs the Anthropocene and the extraplanetary sites characterizing the epoch after. Ecocriticism identifies the bidirectional impact of humans and nature reciprocating in all spatiotemporal locations.

Morton and Haraway

For *The End of the Anthropocene*, Buell's *The Future of Environmental Criticism* acts as a transitional space connecting the formative ecocritical approach with the field's most recent offerings. Buell's book ruminates on areas of thought ecocriticism will encounter in its future use. After Buell, modelling the entangled relations of an ecosystem is distinctly and deliberately complex. Comfortably moving in and utilizing the ecocritical tangle are Timothy Morton's *The Ecological Thought* and Donna Haraway's *Staying with the Trouble*. Morton imagines an ecological framework of thought as a mesh of interconnection and Haraway configures stringy and tentacular pathways. While less linear than the ecocritical connections Buell traces from Marx and Williams, Morton and Haraway are much more entangled in the present web of the Anthropocene. Broadly, Morton and Haraway both trace a framework of interconnectivity for ecoethical preservation of the planet. These theorists initiate an ecocritical continuum for conceptualizing the end of this epoch and the beginning of the next. *The End of the Anthropocene* comes after these texts.

For Morton and Haraway, ecocritical entanglement happens along ecological connections between humans and other organisms in addition to the environment prioritized in earlier ecocriticism. Morton imagines "a practice and process of becoming fully aware of how human beings are connected with other beings" (Morton 2010, 7), and Haraway describes making "kin in lines of inventive connection as a practice of learning to live and die well with each other in a thick present" (Haraway 2016, 1). By their mesh and string figures, Morton and Haraway frame ethical responses for undoing the doubling in the Anthropocene by recognizing the bonds between organisms. Their approaches favor an enhancement of human ethos through recognizing the intricate inextricability of all things within the planetary ecology. *The End of the Anthropocene* treats the entangled environmental construct of these frameworks as a conceptual baseline. Parsing and advancing them with the less abstract application of animal tracklines, ecometonymies expressed in biotic relations are lines of kinship between humans, bacteria, and planetary soil.

The synonymity in Morton and Haraway's perspectives breaks down along temporal lines. Morton is at home in large temporal swaths while Haraway needs little more than the perpetual present. Morton's ecological thought "has to do with concepts of space and time" (Morton 2010, 2) such that Morton asserts thinking through a total connectivity is a process only capable for modern humans. This thinking manifests most clearly as hyperobjects, those things that last well beyond human temporal perspectives. Morton even briefly interrogates the boundaries of environment as perhaps extending to the galaxy (Morton 2010, 10). *The End of the Anthropocene* extends much further, recognizing the universal ecosystem and spatiotemporal nature, and gathering more than "a shadow from the future" (Morton 2010, 2) to witness the very sort of action that ends the Anthropocene and ushers in the next epoch. *The End of the Anthropocene* spends much of its time deciphering the character of the future and its manifestation in the present. A flexible temporal imagination is helpful for reading the rest of this book.

Haraway is more interested in planetary soil and likely regards extraplanetary wanderings as distractions or downright destructive. To remedy the Anthropocene's biotic dissonances, Haraway explains that "Staying with the trouble does not require such a relationship to times called the future. In fact staying with the trouble requires learning to be truly present, not as a vanishing pivot between awful or Edenic pasts and apocalyptic or salvific futures, but as mortal critters entwined in myriad unfinished configurations of places, times, matters, meanings" (Haraway 2016, 1). Haraway constitutes an ethic of biotic relations, showing that biotic process in all its interactions and outputs is an ecoethic that would stifle the Anthropocene's doubling by making humans indistinct kin of any other biotic being. *Staying with the Trouble* details how our species maneuvers the problems of the Anthropocene by prioritizing our entanglement in earthly biotic relations.

To this end, Haraway constructs an integral equation that extrapolates with paradoxical incalculability the whole Earth system. Haraway names the thing made by the equation Terrapolis, that community of all organisms repairing anthropogenic damage by living and dying well together. Critters reside in Terrapolis, "beings of the mud more than the sky, but stars too shine in Terrapolis" (Haraway 2016, 11–12), and by this image I diverge from Haraway. Unlike the critters, we will not just stay on Earth to entrench ourselves in the unfurling ecological consequences—even if we should. Our species already lives beyond the boundaries of this planet, and many intend to inhabit others. We will live further, and earthly biotic connections travel with us into the universe. By this, staying with the trouble as entrenching ourselves in proximal relations with the Earth, seems less likely to be the process by which we procure an ecoethic for living and dying well. I do intend this book to find a path toward such an ecoethic of interrelation even while inhabiting extraplanetary locations.

While the stars shining on Earth certainly refers to a clarity and cleanness granted by an environmental ethos, the stars are also locative.[16] *The End of the Anthropocene* notes how extraplanetary landscapes, on Earth and on Mars, locate a being in the immediate environs of a planet and proceeds to extend that being into those soils by tracks and bacterial kin. Natural navigation entangles stars, planets, dirt, and beings into a biotic identity. Invoking such extraplanetary biotic identities binds Earth and the wilds beyond to recognize the universal ecosystem within which biotic identity is so truly "mathematical and fleshly" (Haraway 2016, 120)—describing extraplanetary wilds is so entangled with physics. For this book, *Staying with the Trouble* is an ecoethic for anywhere produced by invoking the pragmatic futurity in hard science fiction for locating the integrative ecoethical ideal. *The End of the Anthropocene* extends the lines of Morton and Haraway beyond the Anthropocene to approach such ecoethical frameworks.

Toward an Ecoethic

The End of the Anthropocene finds predictive ability in ecocriticism's assessment of the relation between humans and nature when tracing the broad strokes of ecocriticism from Marx and Williams to Morton and Haraway, with Buell as pivotal keystone. In many ways, *The End of the Anthropocene* does an end-around on the present activism of the environmental humanities. At the macroscale, the field works to combat myriad iterations of the climate crisis by such efforts as interrogating the ecological expression of the systemic divisions of the global north and south (Levander and Mgnolo 2011), moderating geoengineering efforts (Morrow and Svoboda 2016), and understanding the increase in global conflict as related to climate (Mach et al. 2019). At more local levels, the environmental humanities seek to secure agency and voice for those impacted by the slow violence of local ecological disruptions that manifest along classic sites of inequity such as race, gender, and income (Nixon 2011). If I were tracing such notions here, I would position Rachel Carson's *Silent Spring* with Marx and Williams and Rob Nixon's *Slow Violence and the Environmentalism of the Poor* with Morton and Haraway. This analysis, though, has little direct relationship to the activism and the social present to which the environmental humanities is currently applied. Theoretically situating this book both before and after is only possible by my trust in the efforts of my colleagues in the environmental humanities to succeed at the current efforts to mediate and realize a positive ecoethic for global environmental stabilization and justice.

The End of the Anthropocene determines the character of the late-stage Anthropocene, the inheritor of our swampy present. Whatever ecoethic we settle upon determines how we control our relation to nature in the late-stage Anthropocene. If there is a deliberate tie to the current moment beyond the standard reflection of the present in futurist texts, it would be in response to Haraway's call: "I think our job is to make the Anthropocene as short/thin as possible and to cultivate with each other in every way imaginable epochs to come that can replenish refuge" (Haraway 2016, 100). When this book interacts with the present, it is to that end. The late-stage Anthropocene begins in the moment we decide the ecoethic that manifests our controlled influence on planetary ecology. Knowing this allows the environmental humanities to codify its efforts with a unifying ecoethic.

Extrication from these exigent efforts affords *The End of the Anthropocene* a vantage from which to posit access points toward a codified ecoethic. Deploying ecocriticism in its literary methodology accesses the cycle of impact by which culture and its texts reciprocally conjure each other. Interrogating the Anthropocene by ecocriticism conjures a perspective by which the imagined realities that have always framed our moral and ethical positions iterate an ecoethic. *The End of the Anthropocene* establishes how we read—that is, how we see and perceive—when we accept the Anthropocene as our geological and cultural location. How might we approach becoming organisms whose most abstract agential act, speech, is an act of perpetual maintenance of ecosystems? Literature has already conjured such beings, and this book interrogates the processes by which we might approach such a becoming by thinking ecocritically about literature with the sciences of indigenous and first peoples and theoretical physics. The ecocritical imagination can recognize such beings and we must aggressively seek to become them.

If we cannot codify an ecoethic that approaches speaking as nature, the humanities will again find itself passed over by a capitalist system favoring technologies of production. Once we decide our path, the Anthropocene enters its endgame. Interrogating the late-stage Anthropocene, *The End of the Anthropocene* reads literary models of the shift from the Anthropocene to the following epoch. The category of the act is firm, although the character of it is undecided until the late-stage Anthropocene. How we act in the late-stage Anthropocene, which ecoethic we embody, will determine the character of the next age. The Anthropocene gets shorter, but not until the end of the next epoch will we get another chance to be better related to wild nature. The late-stage Anthropocene is a transitional space opening into the next epoch. From within, it feels roiling and murky because we are facing the moment in which we will control the anthroposphere. We need ethical stability and certainty that our actions will produce sustainable ecological relation to the planet. For this, I offer the end of the Anthropocene.

THREE EXPANSIONS IN ECOCRITICISM

The End of the Anthropocene deploys ecocriticism to reveal the undercurrents along which the Anthropocene unfolds. Tracing the character of the Anthropocene ecocritically, as it is primarily a relationship between humans and nature, realizes the kind of act that initiates its final stage, the kind of act ending the epoch, and the biotic relations afterward. To assess and affect the future, ecocriticism must initiate three expansions: extend our view to the end limit of the Anthropocene, recognize the Earth system as a patch of the universal ecosystem, and expand ecocritics' scientific literacy to include theoretical physics, astrophysics, and an approach to mathematics. With these expansions, ecocriticism interrogates new texts with a novel methodology to become a predictive theory for codifying the extension of humans and nature into the next epoch. In *The End of the Anthropocene*, ecocriticism becomes a paradigmatic technology examining and shaping the next geological and cultural age.

The Anthropocene's Limit

The first action of this book considers the characteristics and trajectory of the Anthropocene to examine the epoch at its close—termed here, simply, the late-stage Anthropocene. While discussions of the Anthropocene's catalyzing moment are fascinating, especially when curated or filtered by the environmental humanities, this book is not overly concerned with the epoch's initiating circumstances.[17] Precise borders of geological ages are manufactured for comfortable perspectives and productive action but are always instead sprawling transitional sites, fluid contact points that adjoin arbitrary temporal areas that are never independent anyway. Choosing precisely when the epoch begins is retrospective and ultimately unnecessary for characterizing the Anthropocene. Knowing that the anthroposphere is initiated by events ranging from agriculture to the detonation of the atomic bomb is more than enough to understand its character. The Anthropocene is not overly hard to examine, and its beginning boundary is not the limit I am concerned with anyway.

To consider the end of the Anthropocene, we first constitute the core trait of the age: impact on planetary ecology separate from authoritative control. Exemplified by agriculture and animal domestication, importation, industrialization, burning of fossil fuels, deforestation, getting plastic and Styrofoam everywhere, nuclear decay, disposal, and detonation, extinctions, and other self-same anthropogenic actions, human impact on planetary ecology is apparent and persistent. Peculiar, though, is that none of these actions disrupt the planet or even local ecologies malevolently. There was naivete and arrogance, and there may be willful blindness or evil in these acts now, but many on the list began long before humans could conceive of global environmental disruption, of mattering that much, of being able to impact everything, of creating this anthroposphere that modifies the processes of the other great forces of nature (McKibben 2006; Ellis 2018, 14). This idea is by no means redemptive; it seeks only to characterize human impact on planetary ecology in the Anthropocene in order to recognize the nature of its close.

Now, though, we are somewhere transitional, approaching if not already within the late-stage Anthropocene, where *impact* turns to *control*. Again, deciding the precise initiating moment is unnecessary and stymies the foresight enabled by tracing the epoch's character. Organisms existing in the Anthropocene are biologically adapted to the ecology of the Holocene, the previous epoch, and our planet may no longer offer the ecological or molecular resources required to sustain those organisms. Organisms are biotically relating to an unprecedented environment, a planetary ecology that no longer secures their health and survival. Responding to this strange doubling, our planetary impact is also characterized by actions such as conservation, reforestation, pushes for renewable energy, reintroduction of native species, combating invasive species, genetic modification, and de-extinction. Such efforts reflect a positive ethos for living with and on our planet, even if they do not actually correct the terrifying anthropogenic list above. In the late-stage Anthropocene, these actions illuminate human relation to Earth turning from environmental *impact* to *control*. We are moving away from the unintentional planetary impact of agriculture, industrialization, and their ilk and onto the deliberate shaping of ecological contours. The twenty-first century will see our ecological impact characterized by unprecedented command over Earth's ecology to secure our survival.

We will exert such control because we always have. Language and culture allowed us to transmit potent survival practice without the generational restrictions of coding it into our DNA. Since language, our most powerful invention is the scientific method, and its application to planetary ecology revealed the anthroposphere. While science, at its core, is merely witnessing the pattern of physical existence, we seem to adamantly desire its application to our civilized circumstances. Mary Shelley and Erle C. Ellis—regarding, respectively, artificially synthesizing humans and the Anthropocene—characterize science as the Promethean taking of human survival from an overarching supersystem. This is not even uniquely human. In every organism, information creates opportunity to alternatively relate, to survive or procreate better. It worked, do it again. Humans just do this with unprecedented rapidity and impact. Unintentional as it may have been, "In the Anthropocene, humans are put back into a central role on Earth, as planet shapers" (Ellis 2018, 6). The late-stage Anthropocene identifies a shift in the character of this shaping. Identifying the undercurrent of the Anthropocene and its late stage, the scope of this book extends to the end of the Anthropocene to investigate what sort of act ends an epoch.

The Universal Ecosystem

We generally only consider our personal relation to the universe when something from it comes to us or when we leave the planet. Casting off these anthropocentric concepts of environment scales our ecosystem, and thus our biotic relation, to the total universe.[18] In doing so, the blue planet becomes a small patch related to a universal environment through collisions, solar and lunar proximity, and other chaotic circumstances—which it always was, and which life always depended upon. Earthly organisms have always been formed in relation to the abiotic, nonorganic, circumstances that contour the planet.

Generally unacknowledged, objects hurtling through the universe and the planet's distance from our particular type of sun have always determined what forms on Earth. The dinosaurs went extinct when an asteroid impact kicked up so much particulate dust that photosynthesis was distinctly hindered, leaving only small organisms capable of survival (Alvarez et al. 1980). The age of mammals follows. At all earthly scales, from the planetary ecology to the microbiome, the sun is formative and fundamental, illuminating biotic identity's unique extension from an extraplanetary environment. Jack Forbes, as reported by Kathleen Dean Moore, explains that "You could cut off my hand and I would still live . . . You could take out my eyes, and I would still live . . . But take away the air, and I die. Take away the sun and I die. Take away the plants and the animals, and I die. So why would I think my body is more a part of me than the sun and the earth?" (Moore 2005, 58). Biotic identity—what Buell calls "the environment-constructed body" (Buell 2005, 23)—is that which references the biotic and abiotic factors of its creation in relation to nature. Forbes's perspective speaks to the intricate physical expression of biotic relations that care little for the contours of the body, even as it creates them. Biotic identity is about the processes of interrelation to our biotic home, the environs within which we evolved. These connections disintegrate elsewhere in the universal ecosystem and biotic identity is inversely detailed by absence and non-nourishing alternatives. While asteroids and solar radiation are extraplanetary, they alone do not necessarily justify invoking the universe when we are so unlikely to experience whatever exists beyond our galaxy.

Gravity is the most constant and pervasive universal contour experienced in the various planetary ecologies. Our biotic relation to gravity becomes more apparent as we continue our extraplanetary expansion—astronauts experience all sorts of atrophies and physiological effects that just do not happen in Earth's gravity (Garrett-Bakelman 2019; NASA 2021)—although the relation between gravity and the Earth system was apparent to Darwin. Perhaps the most significant human thought since the scientific method, in 1859 Charles Darwin publishes the theory of evolution by natural selection in *On the Origin of Species*. He concludes the tome with a rumination:

> It is interesting to contemplate an entangled bank, clothed with many plants of many kinds, with birds singing on the bushes, with various insects flitting about, and with worms crawling through the damp earth, and to reflect that these elaborately constructed forms, so different from each other, and dependent on each other in so complex a manner, have all been produced by laws acting around us . . . There is grandeur in this view of life, with its several powers, having been originally breathed into a few forms or into one; and that whilst this planet has gone cycling on according to the fixed law of gravity, from so simple a beginning endless forms most beautiful and most wonderful have been, and are being, evolved. (Darwin 2003, 397–398)

The planet perpetuates by gravity, "cycling on." Darwin traces from the earth of a bank the interactions of plants and animals that enact natural selection, to find it seated in Earth's gravity. The palindromic image illustrates the undercurrent of gravity, which we experience constantly not as a feature of Earth but as a feature of the universe that contours the spatiotemporal fabric of reality.

Scientific Literacy

Innately scientific, extending ecocriticism's scientific literacy is the most logical of the three expansions I propose and would occur first if the field was left alone. Second-wave ecocriticism emerges by invoking ecology in assessing human relation to nature and is instilled with a strict care for the preservations of scientific terminology. This care prevents the environmental humanities doing damage to the scientific authority of eco-terms through misunderstanding the sciences or by metaphoric approximation (Buell 2005, 19). The environmental humanities are almost exclusively concerned with the Anthropocene, first imagined as a geological age by atmospheric chemist Paul Crutzen at a conference in 2000 (Carey 2016; Ellis 2018, 1; Schwaegerl 2021). We may even find Crutzen's frustrated exclamation that we are in the Anthropocene to be the most important human notion since *On the Origin of Species*. Recognition of the natural history of biotic relations meant by "Anthropocene" reissues human perceptions of the interrelation of species and their environs.

The most logical expansion of ecocriticism is also the most demanding. Physics is far less casually related to, perhaps even less intuitive than, biological nature. The throughlines between the two sciences are clear, but concepts like the sun's radiation and gravity are fairly intuitive and relatively constant. As humanity expands into environs beyond Earth's orbit though, we increasingly need astro- and theoretical physics, perhaps even orbital mechanics and other applications of mathematics, to conceptualize the biotic relations of humans and novel environs. *The End of the Anthropocene* concludes by tracing gravity and entropy as they frame biotic identity in the next geological age. Gravity becomes an abiotic feature that informs bodily ability and entropy details physical relations to the universal ecosystem. More dramatic than even the Anthropocene's doubling, the human body becomes a disintegrated Earthly construct in the next epoch. These three expansions allow the new ecocritical method to assess a biotic earthly body in extra-planetary environs. That reciprocating ecocritical action imbues ecocriticism with the predictive ability needed to witness the end of the Anthropocene and the beginning of the Astropocene—when the Earth is not explicitly linked to being human.

THE ASTROPOCENE

Something occurs and sends a deep ripple through the structures that inform natural selection, those environmental contours by which nature selects traits that promote survival. Photosynthesizing organisms cycle gases and oxygenate the planet. An asteroid impacts, scattering enough dust into the atmosphere to block out solar rays. The actions and creations of a bipedal species become a new force shaping the planetary system. The next two chapters bring us through the late stage of our current epoch to witness an end to the Anthropocene—an iteration of it anyway. (Predicting the future is hard.) What is apparent is the categorical relation of humans and nature that initiated and characterizes the Anthropocene. That is clear, and so that which ends it is also clear. The Astropocene is when our species lives and dies within extraplanetary ecosystems, on other celestial orbs or in the vacuous environs between. When our biotic self is directly sustained by the universal ecosystem, the Astropocene begins. That new epoch reciprocally alters the planetary landscape in such a way that one must pause the story of Earth if she is to tell the story of terrans on Mars or elsewhere. The same way, if you could track an asteroid in such a manner, you would tell the story of the dinosaurs by telling the story of whatever that cosmic shard once was. The Astropocene is a novel biotic human relation to the universal ecosystem.

As the Anthropocene makes clear, geological ages are also cultural ages. The Anthropocene is not currently an official geological age. Unless we go extinct sometime soon, making the whole naming of ages moot, I see no reason to care. Imagine, enough of humanity recognizes our potentially irreparable damage to the planet—maybe the worst thing we ever did—and then balks at naming the epoch accordingly. Mind you, most of the arguments to do so are scientific, and are born of the same logic that devised the scientific method itself. Culture has deemed this the Anthropocene, and it will be another monolithic human relation to nature, this time as the universal ecosystem, that defines the next cultural age. The Astropocene bears a new human paradigm, not only with novel habitable locations but inviting new areas of inquiry, perhaps comparative geology and transplanetary anthropology. The least of which will be the new humanity conjured by events akin to the first human stepping on the moon. The same biotic relation between human and environment that ends the Anthropocene begins the Astropocene.

The word really should be Astrocene; however, the new era retains its ties to the human impact that will certainly govern its expansion. The Astropocene retains the -opo- even as the microbiome paradigm makes "terran" better than "human." The human is tucked into the word because geological nomenclature is a human construct, and because we will be the species spreading out from the blue planet. Perhaps we eject the -opo- if we end up evolutionarily diverging into Martian and Earth humans. Perhaps it is more important than ever that what we do after the Anthropocene retain its human root. (I reckon we eject -opo- if we meet another sentient species out there, or if there is a circumstance in which a species colonizes the human microbiome.) Geological ages gather to them tectonic and climatological loftiness but are additionally expressed by biotic relations to those environs. In that moment when we become a species with no mandatory biotic tie to our home planet, when other planets will not refuse us—the Anthropocene ends. The Astropocene is the

next human epoch, characterized by deliberate integration and biotic relation
with the universal ecosystem.

TRACKLINES

Morton thinks about a mesh. Haraway sees string figures. I think interconnection with tracklines and game trails. A few years back, I took a week-long
class on sign tracking, sign being the indicators of an animal's presence
outside the track, the footprint itself. We were set to tasks that would attune
our eyes to the relationship between the animal we were tracking and the
landscape. The course focused on evidence like plants that were rubbed,
bruised, bent, or broken by a passing being and hairs that were left behind.
We were set to our tasks with the notion that once we could find the chin
hairs of a vole, we were sign tracking at the appropriate level. One of the
interns had some chin hairs plucked from a vole her cat killed. Those are the
only detached vole hairs I have seen. Most of us learned the visual trick well
enough, at least, to find a lot of deer hair.

One of the exercises was to crawl through game trails snaking back
toward the cedar swamp, which are shockingly easy to follow when you see
them from your belly. They round out at the walls and the plant life bends
around the trail. We were in pairs, and we kept diverging at forks, inevitably running into each other again. Amidst these game trails were primary,
secondary, and tertiary runs for animals ranging from mice to deer. We
found lays where an animal recently curled up for an afternoon and abandoned runs that had not been traveled for seasons. Out the other side of the
entwining trails we pushed deeper toward a creek knowing that the water
and the soft soil always make for good tracking. Each time we approached
the water by a game trail we would find crisp tracks in the mud where a
racoon or deer stopped to drink, or a frog exited the water. The exercise is
designed to entrench us in the relations of animals and their environment
and create a baseline of the space so that we can know where the next tracks
are already.

Tracking is about disentangling one single track or sign from the rest of
the environment so that it can be the method by which a tracker witnesses
how the organism biotically expresses and is expressed by the environment.
Each track leads to another and another. The trackline leads to game trails
which dump right into beds and lays, feeding grounds, and water sources.
In tracking, one engages the entire environment by engaging a single track
or sign. Contained in tracking is an odd temporality. A section of trackline
coalesces in the tracker's present, signaling precisely one animal doing one
thing as though it were there, right then. One can, however, consider that

animal in the future, or the tracker is in the past. Often, the idea is to collapse the timelines and witness the animal. Tracking is a process for reading the environment.

You might think of tracklines as a mesh or string figures, but tracks are literal, and later will be shown to metonymically express the ecosystem entire. Tracking is more praxis than theory, at least it is always praxis first. Morton and Haraway do great work in identifying ecological pathways for our murky present, and in many ways I mean their ideas when I say "ecoethic." I turn to tracking as a material praxis for gathering ecoethical frameworks that persist through the late-stage Anthropocene and into the Astropocene, where novel biotic relations challenge and illuminate what it means to be human. The least metaphorical invocation of tracks constituting an ecoethic, although still harnessing literary imagery rather than tracking praxis, Melissa K. Nelson concludes *Traditional Ecological Knowledge* with a call to use traditional ecological knowledge to "put our environmental ethics into action and get back in our tracks by re-rooting to specific landscapes" (2018, 276). Tracking connects modern science and the proto-scientific knowledge of first peoples toward an ecoethic of interrelation. Tracks detail biotic identity in fiction by the same processes as in reality. Tracks even do this elsewhere in the universe. Any environment in which there is something to depress or brush against, where there is dust or other materials to disturb and transfer, tracking details a biotic being in that ecosystem. To track is to traverse the actual interconnections of organisms in their environmental context. Tracking demands one perceive as another organism in immediate biotic circumstances, and anthropocentric interpretations generally hinder accuracy. Tracklines are a great metaphor for reading or ecology, but for *The End of the Anthropocene*, I deploy tracking as ecocritical framework, avoiding metaphorical slippage into symbolic nature, to image the Anthropocene and Astropocene in the tracks and trackers in fiction—on Earth and Mars.

These texts train us to stare at ourselves in and as the planetary environment, to witness our impact and new landscape. Life, this planet, does not end in the Anthropocene, just as it will not return to what it was before. Human life exists in concert with and because of precise biotic relations in precise atmospheric, solar, and even gravitational conditions. Constructions of self, especially in the Anthropocene, must first witness the contours of the biotic self. Recirculating this with ontological factors reconstitutes the human self with a custom ecological ethos. Ontology is reformed and tempered by tracking's biotic perspective. The species will shortly have to make unprecedented decisions about how we will continue to impact planetary ecology. By those actions we enter the late-stage Anthropocene and initiate the very ethos that will end that epoch and begin the Astropocene. We need ecocriticism's locative and predictive abilities. You all need to do some tracking.

NOTES

1. This is not meant to undermine the naturalism and conservation of the individuals caretaking Yellowstone and the national parks themselves, reacting instead to the macroscale of historical, social, and political interactions.

2. Of course, it has acquired metaphorical significance as well. This is the humanities, after all. I simply mean that it does not depend on it to exist and function.

3. Michael E. Mann takes this position up in *The New Climate War* (2021). He pointedly notes that "We should all engage in climate-friendly individual actions" and that such things alone do not alter planetary ecology: "We cannot solve this problem without deep systemic change, and that necessitates governmental action" (97).

4. Adapted from "Biotic Bodies: Being Human on Earth and Mars," a conference paper I presented at ASLE 2019.

5. Brandon Beshears submitted the video to NBC Montana, which posted it to their Facebook page on January 6, 2020 (NBC Montana). "Wolves in Yellowstone National Park." Facebook, January 6, 2020. https://www.facebook.com/watch/?v=5 29581037645891&extid=12Wj54UG2RhwSbAE).

6. I am using "is become" to simulate a simultaneity and middle chronology of "has become" and "is becoming" as Area X is both. Area X is presently itself, constantly developing—not to mention expanding. "Always already" has the same flavor but lacks the ecogothic agency Area X acquires.

7. Further, Donna Haraway's escalating images of human integration in *Staying with the Trouble* vehemently issue the kinship that disregards human/animal/environment divisions. In the final chapter, Haraway culminates with speculative fiction describing (literal) integration of human identity and ontology with animal kin—a potent ideal, if not pragmatic endgame. Michael Mann's *The New Climate War* helps parse the rhetorical power structures in our culture to redirect individual ecoethical action.

8. I also find this schema useful for resolving some of the issues Clark identifies. Often, he seems to synonymize what I separate: ecocriticism and the environmental humanities. If we instead see that environmental humanists invoke ecocriticism, I think we ultimately sort out the issues of the scaling of the field itself. While initiating the environmental humanities, ecocriticism is itself a literary theory that has been potent enough to manifest an entire human relation to art and the world. While ecocriticism is a part of the environmental humanities, it is not synonymous.

9. *Star Trek, Firefly, Battlestar Galactica*, as well as the more recent *Lost in Space* remake all episodically invoke these tropes. The "Lost in Space" season of *Archer* parodically abuses the episodic invocation and general form of these archetypal SF devices. Less episodically bound to subject characters to these SF antagonisms, film series and stand-alone entries in the genre also diligently maintain these SF tropes.

10. For example, the wolfen trophic cascade of Yellowstone is the most iconic image of the biotic relations of predators, prey, and vegetation, impacting the local abiotic contours of soil and hydrography (Beschta and Ripple 2018). Contrastingly, one might trace the destabilizing impact of invasive species on ecosystems. The

introduction of beaver, for instance, in Tierra del Fuego significantly altered hydrologic landscape features and local biomass (Martinez Pastur et al. 2006).

11. Such as Tom Brown Jr.'s *The Science and Art of Tracking* (1999) and Louis Liebenberg's *The Art of Tracking: The Origin of Science* (1990). Further entries in the study prioritize one angle: *Tracking and Reading Sign: A Guide to Mastering the Original Forensic Science* (McDougall 1999); *Tracking and the Art of Seeing: How to Read Animal Tracks and Sign* (Rezendes 2010).

12. I take a literal approach to tracking in this book. If you are interested in the more metaphorical approach where tracks culturally manifest as footprints, carbon and otherwise, Nathaniel A. Rivera's "Better Footprints" (2018) takes up that angle.

13. Mann, Bradley, and Hughes publish their findings detailing the dramatic upswing of average global temperature (1998). Paul Crutzen and Eugene F. Stoermer propose "anthropocene" for the age of human impact but note that selecting a starting point is "somewhat arbitrary" (2000, 18). Christopher Schaberg epitomizes the cultural circulation of the Anthropocene—which of course includes science: "If the Anthropocene is a useful concept, it may be paradoxically so: to serve as a reminder of such ongoing transition, even as it seeks to name a fairly inertial trajectory of impact so that we might transit in another direction, ourselves" (2020, 39); "The Anthropocene: it's a label for a geologic era, a stratum marked by the cumulative movements and productions of humankind. It's meant to call out humans for disproportionate impact on the planet" (Schaberg 2020, 152).

14. Some may feel that such a distinction is unnecessary, and maybe it is, but as our discipline progresses these distinctions will be made. This is almost always useful to students, as it smooths the ultimately nonexistent replacement of first- with second-wave ecocriticism implied by the naming convention. Describing these demarcations is critical for the environmental humanities, before some bureaucratic entity issues them for us.

15. Published in 1978, Rueckert internally notes that "Literature and Ecology: An Experiment in Ecocriticism" is being written in 1976.

16. As a baseline, I maintain Haraway's work to rid us of the god-gazing version of looking to the stars and naming planets after gods. I truly favor the gorgonic and tentacular entangling of beings and environs. Many who seek Mars may be indecipherable from the patriarchal power systems reminiscent of worshipping such a deity. We are going though. And such, travel immediately conjures an experience in which earthly biotic identities interrelate with non-earthly environs.

17. Let the Geological Society sort that out. Although, if I get a say, I lean toward agriculture wrapped up in animal domestication. In part, because it is a moment of control and design that manufactures terrain, flora, and fauna to warp prior environmental manifestations in favor of our biotic relation to natural resources. This is the exact sort of process that will challenge the Anthropocene's singular planetary location. Food production and stabilizing regions for persistent human civilization will be a primary initiative as we spread beyond our planetary boundaries. (It would be odd to say that the Anthropocene continues on Earth and begins on Mars or hasn't yet begun on Mars.) We will bring to other planets our extensive knowledge of how to impact planetary ecology. We will use it to jumpstart our inhabitance. There is a

sort of collation entrenched in the agricultural character of the Anthropocene that also designs our ecological interactions elsewhere. It's very astropic.

18. I do not make distinctions between solar system, galaxy, and universe here for two primary reasons. First, the primary scientific frameworks, that is theoretical physics and astrophysics, describing the expression of the universal ecosystem do not likely falter elsewhere in the universe. Second, our experiential limitations with the extraplanetary and intergalactic locations makes imagining distinctions between them unproductively speculative for this book. I find a fractal self-same image of the universe to be a more than serviceable framework for *The End of the Anthropocene*.

Chapter 2

Biotic World

Opening Cormac McCarthy's *The Crossing* is a scene in which Billy Parham approaches a pack of wolves:

> They were running on the plain harrying the antelope and the antelope moved like phantoms in the snow and circled and wheeled and the dry powder blew about them in the cold moonlight and their breath smoked palely in the cold as if they burned with some inner fire and the wolves twisted and turned and leapt in a silence such that they seemed of another world entire. They moved down the valley and turned and moved far out on the plain until they were the smallest figures in that dim whiteness and then they disappeared. He was very cold. He waited. It was very still. . . . Then he saw them coming. Loping and twisting. Dancing. Tunneling their noses in the snow. Loping and running and rising by twos in a standing dance and running on again. (McCarthy 1994, 4)

The wolves and antelope move in a space revealed by moonlight reflecting off snow and breath, a seeming moonscape, "another world entire." Billy witnesses an uninhibited wild. Billy knows he will find the wolves here, told by their tracks in the sand and in the snow. He knows the wolves' habitual use of this dry creek bed. His access is granted by the ability to read these tracks and the clothes that barely keep him warm, although he does remain fundamentally separate. In their predation, the wolves display the jollity and authority held by no other animal in the American wilds, and in their predation the antelope express an internal fire no less potently manifesting the inherent spirit of this biotic exchange. Unlike the antelope, Billy remains separate from the wolves, cold and watching. The wolves and the herd wink in and out of Billy's sight and his world, blending indecipherably with that otherworld. Billy, on his family's land, his home, witnesses a scene not of himself and without precedent.

These wolves are indistinguishable from nature, extending as and from the potent, primordial, and uninhibited biotic wilderness. They stoke an inner fire in the antelope that seems counter to the harrying attacks on the herd. The scene does not feel like a hunt, confirmed and subverted by the very verb describing the complex and mythologically common imagery of a hunting wolf pack. Absent is that iconic view of the pack's maneuvers to separate the easiest prey from a herd that moves on from the culminating eviscera-tion and soaked and dripping muzzles of panting alphas with lolling tongues. "Harrying" is just uncommon enough to distance the violent persistence it describes. Then, the hunt's climax is physically moved beyond human gaze.

Distant from humans, we and Billy never see the outcome of the pack's primal herding. The wolves' muzzles are not bloody, and their bellies are not full, but neither do they comport themselves as unsuccessful. They revel, loping and dancing regardless the outcome, just as the antelope are regard-lessly invigorated. The wolves play and lope in failure or success and the antelope are secured in failure or success. They dance, alive. Structurally parsing McCarthy's first sentence demands adept attention to the subjects of the comma-less independent clauses if one is not to attribute the inner fire to the wolves. Although, why shouldn't we? At least, why shouldn't we confuse one affirmation of natural existence for the other? Every outcome for every participant is the natural order functioning correctly. The only thing not affirmed is an anthropocentric expectation that the hunter is the one invigorated by internal fire, a belly warmed by freshest meat. Structural pars-ing deconstructs the human tradition of fire in the predator, placing it instead within the prey animal. In an uncompromised ecosystem, a predator secures the future of the herd. The antelope are intensified and secured with wolves as their curators. The antelope, in turn, curate the wolves, sometimes escaping, sometimes injuring them. Each is nature as it expresses nature. We do not see which iteration because all are identical, and we do not contribute, preferring primordial and wild nature over there.

Billy's separation is cold detachment from the otherworld and deeply com-parative for "Man, in character, is more like a wolf . . . than he is any other ani-mal" (Read 1920, 32). Watching the wolves perpetuate, express, and integrate with nature, Billy experiences a jarring impotence. The absence, both in diction and image, of a killing or failed killing of antelope imbues the scene with notes of herding rather than hunting. Especially through human eyes, this opacity suggests there is no dead antelope. In removing that climactic experience, the scene shifts toward ecological curation of the environment. The image is of total biotic success. Billy is disrupted because "Here is an animal capable of killing a man, an animal of legendary endurance and spirit, an animal that embodies marvelous integration with the environment. This is exactly what the frustrated modern hunter would like: the noble qualities imagined; a sense of fitting into the world. The hunter wants to be a wolf" (Lopez 1978, 165–166). Through McCarthy, the wolf frustrates ranchers as it does those hunters whose duty is to cultivate a herd to sustain a species. Ranchers cannot though, and the very action of human maintenance has all but doomed an entire biotic world. No wonder Billy cannot describe what he saw to his younger kin.

How could I not begin again with wolves? We say things like "put in conversation," but here I would treat as identical the opening of *The Crossing* and the narrative of Yellowstone with which I began the previous chapter. As I write this, the National Parks Foundation continues the promotional campaign that adorns Yellowstone with the tagline "Otherworldly on Earth" (National Parks Foundation). The associated photo of the park's landscape deploys alienating color schemes and a difficult-to-grasp scale, rhetorically skewing away from earthly nature to resonate with images of the various scapes of Mars and the wider universe we constantly receive. Conveniently though, Yellowstone is right here. On the Yellowstone website, the animalia comes right on the landscape's heels. Wolves, bears, and buffalo are still central to the Yellowstone experience—not to mention metonymic extensions of that landscape. Witnessing unimpeded biotic relations is dislocating and full of awe, and we love it.

Billy's experience is precisely what visitors expect of Yellowstone. Ranch or park, the land is parceled and bordered and curates biotic relations within, almost exclusively mediated by some force other than the original ecology, although the landscape's curators encourage some level of natural order.[1] All of this demands humans feel extracted from the ecological exchanges. We simultaneously experience desires for closeness and separation of ourselves and the wilds. Billy never shares with anyone that he witnessed this wolfen otherworld, even though his younger brother is awake when he returns. The natural world is at odds with the Parhams' obligations, and Billy continues his separation from an uninhibited nature. The wolves are separate from the Parham claim and play no role in what the family creates. What really was his other option? They are there to settle the land, and, in that intent, chosen before him without his input, he is separate from the primordial biotic exchange of the wolves and antelope.

THE ECOLOGICAL GAZE

The wolves then look at Billy:

> There were seven of them and they passed within twenty feet of where he lay. He could see their almond eyes in the moonlight. He could hear their breath.

He could feel the presence of their knowing that was electric in the air. They
bunched and nuzzled and licked one another. Then they stopped. They stood
with their ears cocked. Some with one forefoot raised to their chest. They were
looking at him. He did not breathe. They did not breathe. They stood. Then they
turned and quietly trotted on. (McCarthy 1994, 4)

The scene's otherworldly quality is compounded as the wolves stop and
return Billy's gaze and move on without care. They extend a gaze riddled
with distancing uncertainty and superior unconcern, giving the impression of
the wolves, of nature assessing the boy. Ecocritically parsing the scene's perspectives reveals various gazes and false gazes circulating within the biotic
relations of the organisms and their environments. Especially applicable
after linking the ranch and national parks by the Anthropocene, tourist gazes
sit akin and opposite those of this interaction and comparatively articulate
Billy's experience as one who feels separate but is already contained within.
The wolves reveal a preceding assessment by nature's gaze, Billy was never
outside their awareness, masked by an anthropocentric hubris. In this scene,
The Crossing conveys the characteristic human dissonance and inclusion in
nature through gazes and biotic relations.

Billy gazes upon the wolfen otherworld as a human separate but evinced
in some manner. Useful as an initial frame, Urry and Larsen's environmental
gaze is most ecocritically minded, a gazing ethos in which the tourist is concerned with ethical human impact on the environment before consuming the
spectacle (Urry and Larsen 2011, 20).[2] Thinking of the underlying character
of the Anthropocene, the distinction between a national park and a ranch is
murky, but gradations in ethics and wildness certainly delineate Yellowstone
and cattle ranches. Both such bordered landscapes constitute relation to the
environment constructed by anthropocentric impact and control. The environmental gaze houses an ethos for preservation contouring how one would
view natural sights. Thus, this gaze seems most ecocritically applicable to the
scene because, in this moment, Billy does not intend to confront or disrupt the
wolves. Still, the talk is of tourist destinations and cattle ranches; capitalism
is present, but where is it not? I would reissue VanderMeer's sentiment: you
might think you have but you have never moved through an uncompromised
ecosystem (VanderMeer 2014a, 32).

That night though Billy may have been closer than any of us. There are
flavors of the spectatorial and reverential gazes—Billy gathers a relatively
small, snapshot-like vision of an ecosystem, and he is deeply affected,
although not invigorated, by the otherworld he witnesses. These gazes
ultimately fall short, not from Billy's lack of formal tourism but through
the scene's lack of human curation. The wolves harrying the antelope
emanate from a transitional space of preceding wilderness running up to

this less-than-settled southwest. *The Crossing*'s wolves are primordial, wildly preceding, the Anthropocene. Yellowstone's wolves are more complicated. These Canadian transplants are indeed wild wolves of primordial stock. Yellowstone's packs indicate both the glitch in wolfen presence and the historic biotic relations. They and the vanished originals are from the same wolves that shaped the ecology of the entire North American wilds. (Wonderfully, we can count on wolves to be wolves.) The action is affirming and vexing. There is no doubt that the reintroduction of apex predators signals an ethos that values pre-Anthropocene environs. Yellowstone, however, remains a microcosm of the Anthropocene.

Human impact and control of the environment in the Anthropocene belies all actions of the Yellowstone wolves. The park's borders, mission, and business constitute the ecosystem; the wolves descend from those reintroduced by human directive; asynchronous state legislation impacts the wolves because their territories do not comply with state borders; federal legislative action is riddled with maneuvers unrelated to the wolves. The environmental gaze abrades against more fervent environmental ethics because it ceases to exist if any human presence could disrupt the location—consider that you cannot visit Methuselah, the oldest living tree of confirmed age. Billy's wolves, like Methuselah, have a primordial essence all national parks seek to contain or duplicate but, in the Anthropocene, can only reference. The parks do their best to speak a word for wild nature, but all current ecosystems pivot on human impact because they include the anthroposphere, even if they are not definitively curated for tourist gazes. When the Parham's arrive, the southwest wilderness still overlaps with ranches, although by that time Yellowstone's wolves have been exterminated. Billy gazes upon revenant wolves in their biotic home and witnesses pre-anthropic nature, otherworldly and primordial.

Billy is forever silenced by the otherworld he witnesses, never reiterating that primordial landscape. The experience is stripped of any remaining tourist trappings as he never asserts his presence. Tracking the wolves on his family's land is likely motivated by the intent to hunt and trap them as they potentially threaten livestock, but the knowledge gleaned from the tracks is used to witness them as they are in the world. Apolitical and stripped of capital's motivation, upon what did he gaze that could command him so completely? A combination of the environmental and anthropological gazes focuses on the historical interactions and trajectory of the scene's species and environs. With this ecological gaze, Billy sees a pack of wolves harrying a herd, a primordial process by which the interrelation of the wolves and antelope invigorates each. This relation constitutes and perpetuates the environment by a superior process preceding civilization, a biotic integration he cannot have even though the ecoethic is conjured within him.

The killing and consuming that sustains the wolves promotes the herd. Billy's gaze collapses ecological history into these wolves:

> Wolves were the driving force behind the evolution of a wide variety of prey species in North America after the last ice age, literally molding the natural world around them. The massive size of the moose, the nimbleness of the white-tailed deer, the uncanny balance of the bighorn sheep—the architect of these and countless other marvels was the wolf. (Blakeslee 2017, 14)

Predator-prey relations shape and invigorate species. The hunting actions of wolf packs culled small moose leaving only the large individuals who were capable of fending off attacks while white-tails became defined by deftly bounding through and over thick brush. The casual ability with which big-horn sheep climb sheer cliff faces boggles the mind. Every facet of the North American ecosystem is conjured, strengthened, maintained, and secured by the biotic relations of wolves, and contained in that statement are the pathways by which one may connect the other ecological relations of the biosphere. Big-horn sheep, the cliff face, the briars white-tailed deer bound over, the flood plains maintained by aspen all describe the wolf. The ecological gaze focuses on one biotic extension of the ecosystem as it implies all others.

The ecological gaze must be carefully deployed, beginning with the objective findings of environmental science. Scientific literacy mitigates logical assertions extending from unsteady frames. However, the ecological gaze is not simply the gaze of environmental science. This is the Anthropocene after all, so throw in humans and culture. Here is the mesh of Morton's ecological thought, which is infinitely all things as they are connected. The ecological thought and the infinitude of access points is wonderfully invigorating and vastly displacing. Witnessing the mesh described by wolves and antelope, Billy's silence suggests an ethos he does not embody. Billy watches apex predators perform a perfect sort of herding that maintains and secures the total ecosystem. Wolves and antelope live and die well together. The ecological gaze witnesses this innate sustainability that Haraway positions for us as an ethos.[3]

Billy witnesses in the wolves an unachievable and tangible, perfect version of himself maintaining nature by the very act of maintaining one's family. The Parhams are no such herders. Billy is not even a steward or warden of the environment. He experiences human digression from any such innate biotic maintenance, instead adding the additional steps of turning cattle to capital and back into resources, further distancing himself from any image of living integrated with the land and his biotic home. Billy sees the otherworld, pathlessly distant from his anthropogenic existence that maintains civilization at the expense of the invigorating exchanges of the wilds. Billy is silenced by the impossibility of the wolf. The wolves and their biotic relations are unable to maintain primordial nature in the Anthropocene. Wolves have been dislocated and eradicated and, now, humans confront the same ecological future. Our hunting and herding seemed wolfen but maintained anthropogenic sites, not those wild places in which we evolved. Humans failed to be wolves, and the actions of wolves now only replicate primordial otherworlds at the behest of anthropos. Perhaps even wolves now fail to be wolves.

ECOMETONYMY AND NATURE'S GAZE

The ecological gaze accesses ecosystems by examining ecometonymies, the environmental parts extending from and as the ecosystem to imply the existence of the larger ecosystem. Ecometonymies chart a single access point's relation to nature, reciprocally constituting and constituted by nature's systems.[4] The wolves and antelope sustain each other, and the existence of one implies the other. Above, Blakeslee's description of North American wolves ecologically gazes at the wolves of Yellowstone and their biotic interactions, tracing speciated expressions of pre-anthropic North America by predator-prey relations. This action is a gaze, so it fervently stares at one organism. Blakeslee's description never moves particularly far from the wolves because the book is a biography of a single Yellowstone wolf. Thus, the above passage does not flesh out the abiotic factors of the landscape that the description implies.

The ecometonymic action traces a path to other species and landscape features. Ecometonymy reveals the landscapes implied in Blakeslee's description of prey species becoming larger or maneuvering adeptly within their environs. Even halting on the cusp of the role terrain plays in predator-prey exchanges, a close-reading quickly reveals forest density, brush height, and cliff faces. Tracing ecometonymies is a potent connective power in which one single wolf expresses the contours of a continental ecosystem. Billy's scene helps strip anthropocentrism from the ecological gaze. Antelope or wolf, the organism describes the predator-prey relationship that shapes the local ecosystem, which we can pursue up to ecology at the planetary scale. Gazing at wolves is to gaze at other animals and plants and landscape features, quickly detailing the entire ecosystem. The ecological gaze is rooted in the humanities; as such, it is governed by philosophical discourse. The ethos of this gaze is maintained by ecometonymies as they scientifically describe organisms living and dying well together. Humans are generally more fascinated by wolves than antelope, but neither predator nor prey is less invigorated than its counterpart, both, in *The Crossing* or reality. Only the human is cold, for he is not living and dying well with any being.

Closing the otherworldly scene, the pack then gazes upon Billy. Ecocritically reading the gaze's distancing unconcern and otherworldly tone refines the possible motivations for the wolves' attention to the gaze of biotic wolves, ecorealistic instead of moral. They do not eat Billy, so the gaze is not predatory, but neither do the wolves see a peer. Wolves are unlikely to reflect upon their ecological relations to humans in the Anthropocene, so any other perceived synonymity would likely mark an interloper needing to be routed for territorial enforcement. Possibly, the wolves are familiar with humans treating them with casual observation, although this runs counter to the aggressive extermination of wolves by ranchers like Billy's father staking a claim and the otherworldly tone of the novel. Instead, Billy does not register. In anthropocentric hubris, one might read this as an ejection from the natural order. As ecocritics, we are especially susceptible to readings condemning acts like the extermination of wolves—especially in a novel published one year before Yellowstone's reintroduction project by an author who reportedly plotted to reintroduce wolves to the southwest (Greenwood 2009, 65). Counterintuitively, reading the scene as a condemnation is dangerously anthropocentric.

The wolfen gaze cannot signal ejection because organisms are inextricable from nature. The Anthropocene marks a geological age in which human impact on the environment—not constant, proactive human maintenance of the anthroposphere—disproportionately influences planetary ecology. The Anthropocene does not end tomorrow if we go extinct today. The anthroposphere is comprised of human-crafted hyperobjects, not humans themselves. The wolfen gaze does not describe a human separation or ejection from nature. We have not even exited the biosphere, now feeling the doubling impact of harmfully altering the planetary ecosystem. The most anthropocentric thought is that we have extracted ourselves from biotic relations to other organisms and abiotic factors, now existing beyond or aside nature. We have constructed an entirely new sphere to the Earth system, but it is still categorically abiotic in its shaping of how planetary and local ecosystems unfold. The wolfen gaze confirms we are still well within nature's full regard.

The fundamental exchanges formulating nature remain unchanged. Organisms succeed or fail according to their physiological abilities in the environment. The wolves and Billy are not biotically separate. We recognize the otherworld as Billy is doubled, witnessing revenant biotic relations in which species were integrated with the landscape and each other. Billy witnesses pre-Anthropocene nature ecometonymically expressed by wolves and antelope; he witnesses the Anthropocene ecometonymically expressed by wolves, antelope, and himself. In the transitional space between the Holocene and Anthropocene, wolves and humans ecometonymically describe both. Both species are physiologically adapted to the Holocene and their struggles and extinction reference the new landscape of the Anthropocene, doubled as though in VanderMeer's *The Southern Reach Trilogy*. Even stripped of anthropocentric bias to epochal demarcations, these epochs are just manifestations of categorical life systems within nature.

Gazes and metonymy each trace back to their projecting container, completing a transitive conjuration of nature's gaze that reasserts the human ejection I describe as anthropocentric hubris. Billy's and the wolves' gazes backtrace directly to the visual organism while organisms express the biotic and abiotic factors of their ecosystem. In the sense that natural selection is the process by which nature assesses organisms, nature gazes. The gaze is a metaphoric process completed by anthropocentric approximation that asserts nature assesses through eyeballs like humans. Even "assesses" may be dangerously approximate. Metonymy, as not metaphor, charts a realistic extension. Ecometonymies potently express metonymic ecorealism as they function to identify biotic identity by ecological relations appearing in nature and in literature that meets second-wave ecocriticism's necessity for realist natural representations—those built from environmental studies' cataloguing of tangible biotic relations. Nature gazes metaphorically, not metonymically. The concept of nature's gaze is again a metaphorical anthropocentric co-opting of nature's formula, leading very quickly to ideas of humans functioning beyond nature instead of organisms biotically related to nature, as always. Nature's gaze equates natural selection with a visual action by a metaphoric process that implies the ability to exit biotic and abiotic ecological relation— as though we could avoid nature's gaze, somehow exiting biotic relation with our environment. Ecometonymies trace the inextricable biotic connections described by environmental science and ecocritical literature.

Maintaining ecocritical tenets of reading the wolves as nonmetaphorical, issuing no value judgments, treats a human as an organism in the environment. If not predatory, territorial, or familiar, then the gaze seems inclusive, though not welcoming. This inclusion precedes the wolfen gaze, and *The Crossing* deprioritizes the very notion of gazing when Billy "feels the presence of their knowing" (McCarthy 1994, 4) prior to witnessing the wolves conclude their care of each other, halt their revel to regard him, and anticlimactically walk off. All parties are aware of each other well before gazes anthropocentrically consume the reader's sensory experience of the image. Just as he is not eaten, Billy may escape this perspectival consumption. The wolves' knowledge of Billy's presence that precedes their gaze travels along innate and primal inclusion in nature. In favoring the scene's gazes, we think anthropocentrically and fail to engage the preceding sensory awareness, which is less tangible. The visual sense subsumes other attentiveness that reveals the biotic relation of the physical self to nature. Anthropocentrically minded, the wolves gaze and so nature gazes upon humans as other.

Ecocritically, Billy notices himself as always already in biotic interrelations that formulate the natural world. The inclusion of Billy and his knowledge of his inclusion precedes the gaze of the wolves. Our attention to the wolfen gaze anthropocentrically consumes that preceding awareness. Understandable, for we are a highly visual organism with deep cultural affinities for eyes, imagery, and gazing. All poetic sensory experience is categorized as "imagery" and blindness describes any misunderstanding. The wolves, though, are not literally nature but are metonymically expressive of nature's constituting formulas, in fiction and environmental science. The wolves and antelope are metonymic extensions of nature aware of Billy for they express the entwining interaction of biotic organisms in the immediate environment. Billy too is an ecometonymy, but of the Anthropocene, which is just a container signifying the characteristics of nature derived from the same old interactive formulas. The wolves' awareness as that of nature has already included Billy, for he has never been extricated from it. Instead, this gaze is but an ecometonymical expression of nature that has drawn Billy's recognition of himself in the Earth system.

Humans contribute to the character of the natural world with unprecedented potency, but we have not altered the formula. Organisms respond to their environment and the environment responds to the organisms. The difference between Billy and the wolves is best described in geological epochs as they are determined by the average consistency of planetary ecology. An organism either succeeds or fails, but the failure is always biotic. A dying tree never undermines the ecosystem. The tree is just doing tree stuff. A fox too weak or near-sighted to survive just does fox things until it dies, never altering fox speciation in any significant way. In either case, the ecosystem continues.

Humans have initiated a shift in nature that is unprecedented among animals, but the underlying means by which nature constructs itself are unchanged. The innate maintenance of the ecosystem within which a species evolved may be the only difference between humans and any other animals. We need an ethic to prevent our impact from being an extinction event while no other species needs philosophy to ensure the Earth system. (We need this to be the Anthropocene, not the mass extinction ending the Holocene.) We seek that ethos and the question becomes how to live and die well in the Anthropocene's environments. How might we become ecometonymies of a balanced Earth system instead of ecometonymies of the sixth great extinction? Or, to gaze directly at ecocriticism's most destabilizing axiom, how might we speak as nature?

WITNESSING NATURE: THE HUMAN ANIMAL

In the Anthropocene, Billy's father interacts with the landscape with an ancient and alien, even mystical feel:

Crouched in the broken shadow with the sun at his back and holding the trap at
eyelevel against the morning sky he looked to be truing some older, some subtler instrument. Astrolabe or sextant. Like a man bent at fixing himself someway
in the world. Bent on trying by arc or chord the space between his being and the
world that was. If there be such a space. If it be knowable. (McCarthy 1994, 22)

Setting a trap, Parham gathers to himself a relation to the biotic, formerly
wild world akin to the sextant's ability to gather a singular location in this
planet's uncharted places by the unerring patterns of the galaxy. Utilizing
traps to respond to a wolf arriving on his ranch ten years after Billy witnesses
the wolfen otherworld invokes Parham's most biotically integrated self, when
he secured his claim to the land by settling the wild. Parham's life and the
survival of his family once depended on altering predator-prey relations to
stabilize his economic self in civilized food production—seemingly undermined but not unraveled by his son's connection to the wolfen otherworld.
He helps institute and maintain the Anthropocene in the southwest. Parham's
interaction with the environment is indivisible from anthropogenic civilization, and is never utilized, if there be such a use, to maintain the environment.

Parham fails to trap the wolf as she spends her nights uncovering, without triggering, each set. Parham's failure runs counter to his tracking ability
which readily provides nuanced detail of the wolf's biotic self. The location
of urine describes the wolf's sex and unnarrated sign reveal to Parham she is
pregnant (McCarthy 1994, 16, 23). Billy succeeds though, trapping the wolf
with a set built under the prior day's spent lunch fire. The wolf has taken to
rooting out the food scraps vaqueros have discarded into the ashes, described
to have begun living by new protocols (McCarthy 1994, 25). I can think of
no better Anthropo-scene than formerly Promethean fire used by cowboys for
comfort and maybe reheating a packed meal having unconsumed food dirty-
ing a plate scraped onto its embers and ashes. Parham's failure and Billy's
success describes Parham attempting to trap a pre-Anthropcene wolf, and
Billy seeing a wolf in the Anthropocene. This wolf survived because she has
adapted her otherworldly self to the current epoch that now extends over the
entire southwest and extinguishes the wolfen otherworld.

The Parhams live closer to uninhibited nature than we normally do, even
though Parham is attuned to a pre-Anthropocene environment while Billy
engages the wolf in the current anthropogenic world. Their ranch, or any
modern incarnation, maintains a larger portion of that ecosystem's historic
biotic and abiotic characteristics than rural areas just outside cities—even if
we include red-tailed hawks riding highway thermals or ants in our yards.
Billy and Parham see the biotic world as we do not, be it an otherworldly wolf
dance or a wolf's biotic identity in her tracks. Tragically and usually, those
with the ability to read the environment gather their vision for the progress

of civilization, not nature. Billy's effort to return the wolf to a wolfen otherworld fails by the pervasiveness of the Anthropocene, despite the positive
ecoethic he enacts.

Disconnect from the wolfen otherworld is borne of the same Anthropocene
stuff that makes the paradox of the ecocritic more defeating as identified by
Lawrence Buell:

> One can speak as an environmentalist, one can "speak a word for Nature, for
> absolute freedom and wildness," as Thoreau did, but self-evidently no human
> can speak *as* the environment, *as* nature, *as* a nonhuman animal . . . At most we
> can attempt to speak from the standpoint of understanding humans to be part of
> what Aldo Leopold called "the biotic community"—attempt, that is, to speak
> in cognizance of human being as ecologically or environmentally embedded.
> (Buell 2005, 7–8)

We are firmly in the Anthropocene though, where even the actions that make
us feel connected to nature orbit civilization, unlike nature itself.[5] Hugely
problematic, the inability to speak as nature is perpetuated by humans connecting with nature as othered by the Anthropocene. Certainly, it is easier to
speak as the city, as citizen, than as an organism. Connectivity to nature is filtered by civilized otherness keeping nature "over there." The disconnect from
natural perspectives possibly even produces symptoms like being able to
deny climate change, the very circumstance of our ecosystem. Billy traps the
wolf and then brings her home to Mexico. This relocation fails because there
is no wild left. The wolf arrives at the ranch for this very reason. In Mexico,
Billy shoots her instead of permitting her figurative and literal consumption
in a dog-fighting pit. Even this pregnant wolf, using its Holocene adaptations
to continue in the Anthropocene, is killed by the closest thing *The Crossing*
can find to a naturalist. How might we speak for wild nature consumed and
controlled by the Anthropocene?

TRACKING THE BIOTIC HOME

Speaking as nature necessitates ardent and total integration with one's biotic
home. What that looks like is difficult to imagine, though is this not precisely
what fiction is for? In *An Imaginary Life* by David Malouf, a fictional Ovid-
in-exile details his interactions with a feral child who grew up among deer.
The Child feels no cold on his naked skin, which seemingly absorbed the soil
and its tones, and can keep pace with running deer as he moves through the
wilderness beyond the bounds of the Roman Empire (Malouf 1978, 114, 77,
48). The gait in his tracks describes his speed and place among deer. After

capturing, befriending, and exchanging knowledge with the Child, Ovid recounts their walks through marshy wilds:

> All this world is alive for him. It is his sphere of knowledge . . . another language whose hieroglyphs he can interpret and read. It is his consciousness that he leads me through . . . It flickers all around us: it is water swamps, grass clumps, logs, branches; it is crowded with a thousand changing forms that shrill and sing and rattle and buzz, and must be in his mind . . . Only for him it is a visible world he can walk through, that has its weathers and its seasons, its cycle of lives. He leads me into his consciousness and it is there underfoot and all about me. (Malouf 1978, 93–94)

The Child is inseparable from nature, existing with no distinguishing border between his consciousness and his environment. No internal/external binary needs or is deconstructing because the Child's history contains no such ontological or bodily excision from wilderness. As in dreams, which are so important to the novel, the environment simultaneously contains the ontological self and is synonymous with consciousness; the difference being that nature, unlike a dreamscape construct, is a tangible and preceding location. The Child's consciousness is the total, external environment and all biotic relations within. His thinking is indistinguishable from biotic wilderness. The environment as consciousness allows the Child to thrive in environs foundationally opposite civilization. Contrary to the relationship of civilized individuals and wild nature, the Child is never cold, without food, or overwhelmingly problematized by apex predators. His body, like his consciousness, is indistinguishable from nature. The Child is a fully integrated human animal. He must be capable of speaking as nature.

As nature is his thought, so too is nature his voice. On their walks, the Child teaches Ovid about the wilderness and its inhabitants by producing the vocalizations of animals they encounter. Ovid explains, "It is as if each of the various bird species . . . had their life in him and could be drawn out on the breath between his lips; as if he had some entrance to their mysterious comings and goings among the grasses, or had been with them to the bottom of the river" (Malouf 1978, 91). His consciousness is the environment, and, in this way, he has been with animals in their nonhuman spaces and has himself embodied those places. Beyond this metaphysical explanation, the Child has likely stood in those places, literally. He travels game trails, swims in rivers and swamps, burrows in dirt, and hides from predators in scrub brush and trees. Such a life grants him integration of his consciousness and the environment. His ability to speak as nature would at some point demand understanding animals' material lives in biotic relation with tangible nature.

Additionally, the Child's mind thinks as and with all creatures, so it follows that his oral communications are theirs—though, this should not be confused with the act of calling in an animal. Ovid explains that "In imitating birds, he is not, like our mimics, copying something that is outside him. . . . He is being the bird" (Malouf 1978, 92), for as he makes the sound:

> He stands with his feet apart, hands on hips . . . and his lips contort, his features strain to become those of the bird he is mimicking, to become beak, crest, wattles, as out of his body he produces the absolute voice of the creature, and surely, in entering into the mysterious life of its language, becomes, for a moment, the creature itself . . . he seems miraculously transformed. (Malouf 1978, 90)

The Child has created a pathway from consciousness to body via voice. The human body is not evolved to innately make the sounds of other species, but, by contorting his body, the Child enacts the shape both by making and to make a sound that is distinctly *the* animal. (He begins to learn our language by feeling Ovid's throat.) The Child uses anatomy and sound to conjure each other, which they do, extending himself to the evolutionary pathways through which the environment formulates species.

Any sound in nature articulates the initiating organism and its biotic circumstance in the landscape. A twig snapping indicates an organism suddenly failing to walk silently, although it seems to have done a great job in getting so close. This causes birds to pause and assess the sound, and if that snap came from me instead of a deer, their alarm ripples outward, taken up by other birds, squirrels, chipmunks, and reacted to by the nearby fauna. Depending on its assessment of my intent, the bird might even fly from tree to tree above me, alarming, marking my movement through the space, and warning an entire area. Sound extends in concentric spheres alerting all organisms, not just those of the initiating species. Tracing sound to realize the environment it resonates through and from is an historic ecocritical pursuit. Considering how sound construes the landscape in *The Machine in the Garden*, Leo Marx invokes Thoreau's "Sound" chapter in *Walden*:

> Now [Thoreau] shifts to sounds, "the language with which all things and events speak without metaphor, which alone is copious and standard." The implication is that he is turning from the conventional language of art to the spontaneous language of nature. What concerns him is the hope of making the word one with the thing, the notion that the naked fact of sensation, if described with sufficient precision, can be made to yield its secret—it's absolute meaning. (Marx 1964, 248)

Marx identifies this "pastoral interlude" as Thoreau's attempted movement toward the nonmetaphorical, true expression of biotic circumstance—which is the ecometonymic pathway from sound to organism to environment. Marx and Malouf describe while Thoreau and Ovid seek the perspective of the biotic world that grants the Child's connective becoming, his identity as one who can speak as nature.

In *An Imaginary Life*, Ovid even wonders if he might learn and then write poetry in the language of spiders, resonant with Thoreau seeking to understand and convey the thing with the sound. Poetry's fundamental tenets of sound and metaphor describe the problem of humans becoming like the Child. Poetry's use of sound and word to constitute lived experience describes the human capacity to understand true absolute sound-meaning, while the inherent metaphoric relation makes ecometonymic traversal impossible. No wonder Malouf's receiver is a poet, and no wonder the sound of the train shortly draws the attention of Thoreau and Marx. (This is the Anthropocene after all.) The Child manifests and inspires the perfect connectivity Thoreau was after, a pure expression of integration with uninhibited wilds, allowing the Child to physically pass himself through and become all ecometonymic expressions of the ecosystem.

This is not to say that speaking as nature is the ability to speak as other animals. Rather, acquisition of such languages as they describe bodies in the landscape is a precursor to speaking as nature.[6] Resonating with Haraway's chthonic ones—"beings of the earth, both ancient and up-to-the-minute . . . replete with tentacles, feelers, digits, cords, whiptails, spider legs, and very unruly hair" (2016, 2)—the Child's consciousness is a natural chimera amalgamating the landscape with its inhabitants, and their shape, and their voice, and their food, and their impact on the landscape, and all other ecological interrelations witnessed by an ecological gaze. All that is biotic, is the Child. Through this, the Child even seems to transcend Ovid's temporality, simultaneously appearing in his past and present. The Child is, as nature, unbounded from time. With his perspective, the Child's bodily life is so antithetical to Ovid's that he fractures the binding notions relating Ovid to spacetime. *An Imaginary Life* closes with Ovid's perspectival integration with the wild which deteriorates the difference between ecology at various scales and the material integration through the biotic cycling of death and decay.

The heart of the matter is finding a language, a system of signs, that allows humans to think as the Child. We are not the Child, with most of us less equipped to read the landscape than Parham. Ovid gets there though, and what a relief for us. The Child instructs; Malouf instructs. Now, all we need is to learn, unlike Ovid, significantly before our death.[7] Ovid describes the language of the Child, of nature, as hieroglyphic vocabularies akin to his internalized knowledge of poetry, theology, rhetoric, math, science, history, and philosophy. Looking out at the wild scene(ery), the Child knows what resources and threats are where, what is edible, and how to maintain his bodily life. The difficulty of learning this semasiographic language is that, unlike linear and finite hieroglyphic writing, nature is self-entangled and ever circulating. Deriving the grammar and vocabularies is a gargantuan task still incomplete even with the extensive efforts of biology, environmental science, organic chemistry, physics and other such disciplines (with more derivations to come, I'm sure). What signs exist to give us access to the biotic interrelations of planetary life? In answer, I would remember the first proof of the Child's existence: his tracks (Malouf 1978, 48). Even the storytelling before Ovid witnesses those tracks, a story of the Tomis people finding the Child's tracks regularly, is less real than the story the fresh tracks tell—a true story of the Child's body being able to keep pace with deer.

A track is a locus of biotic interrelations describing the confluence of body and environment, while an entire trackline informs on the organism in the eternal motion of nature. Learning to track is learning to read an ecosystem, to think as the Child. The Child's lessons for Ovid are grounded in tracking, locating for him "tracks in the grass and explaining with signs or gestures of his body, or with imitation sounds, which bird or beast it is that has made them" (Malouf 1978, 93). Tracks are literally ingrained in the landscape, describing an animal in its environment as it responds to immediate circumstances. All tracks are proof of an organism in the landscape, a readable moment in the progress of living in the environment. When reading tracks, the Child uses his body and voice to convey which animal was exactly there. We must remember, that when the Child contorts his shape and sound, he becomes the animal who made it—and, as the next section will show, the track, thus the Child, is the exact and individual animal, not merely the species. Tracks are bodies, if one learns to read them. With this language, one approaches thinking as the Child. One approaches being able to speak as nature.

TRACKING TO THE BIOTIC HOME

Parham, Billy, and the Child are each accessing the exact organisms who left tracks, those impressions in the landscape that detail a biotic identity contextualized by the environment and other organisms. This individuating precision is largely mythological in modern culture, epitomized by quintessential tracker archetypes like Aragorn parsing the Hobbits' trail from the tracks of a recent battlefield in Tolkien's *The Two Towers* or how, in McMurtry's sequel to *Lonesome Dove*, Famous Shoes consistently knows the identity of a person by their track as though they were standing in the depression—the latter provides a deeper explication of tracking. Images of tracking in culture are constant: the tracker sees that the track equates to the being who made it. McCarthy's *The Crossing* is a sort of gold standard of the archetypal tracking action, more diligently describing the tracks and how they detail biotic identity. Always though, in fiction or our own environs, tracks connect the tracker to that individual because they describe the precise actions of a being in that environment. Tracks metonymically extend from the organism, describe that being's actions, and enable pursuit and arrival at that individual regardless of any convoluting factors impacting the trackline. But for temporality, the being is standing right in front of the tracker.

In ecocriticism, a track in the landscape is stripped of any metaphorical, thus mythological, identity and is simply a depression in the environment. Without fictional mythos, the track maintains its identifying power and the tracker's formerly mystifying ability to describe the organism remains in place. Tracking performed in fiction produces the same outcome when performed in our real environs. Tracking is a powerful ecocritical motif, for it functions identically in fiction and reality to describe biotic relations. Maintaining ecocritical process to read actual nature in fiction eschews the mythological perspective attached to the tracking action to show that the mystical vision of fictional trackers is, simply, tracking. Tracks issue the absolute existence of an organism, a knowledge inextricable from human evolution.

In a chapter of *Born to Run*, Christopher McDougall details that pre-Stone Age humans deployed tracking as the critical means of hunting herd animals for consistent large sources of protein. McDougall describes that human physiology contains equipment needed only by running animals. The Achilles tendon gives that springy step while the nuchal ligament keeps our head from flopping around (McDougall 2010, 221, 220). We sweat to cool down instead of panting, continuing when other animals overheat. Upright structure affords a more efficient and adaptive breathing rhythm than, say, a cheetah, who scrunches, accordion-like, in exhalation with each step as organs slam forward (McDougall 2010, 216, 223). Humans even have a more efficient stride than horses, covering more distance per step (McDougall 2010, 222). These human features describe a long-distance runner capable of persistence hunting, chasing an animal to death. A human at a jogging pace forces a prey animal to sprint without sufficient rest, hyperthermically killing it (McDougall 2010, 227). Humans develop this hunting method as the last ice age ends and the forests recede. The open plains are a perfect landscape in which to chase animals for miles (McDougall 2010, 227–228).

Louis Liebenberg, a student of mathematics and physics, evidenced this ability, learning persistence hunting while living with the Kalahari Bushmen. The trouble with any strong runner just going out and chasing down herd animals is their tendency to become indistinguishably part of the herd or environment when chased. The Bushmen taught Liebenberg "to look at piles of dung and distinguish which droppings come from which animal; intestines, he discovered, have ridges and grooves that leave unique patterns on feces. Learn to tell them apart, and you can single out a zebra from an exploding herd and track it for days by its distinctive droppings" (McDougall 2010, 235). This is the level of tracking ability needed to persistence hunt. Liebenberg explains that "the art of tracking requires fundamentally the same intellectual abilities as modern physics" (Liebenberg 2013, 17). Even further, discovering such knowledge demands the fundamental scientific perspective and process, a sort of proto-praxis. Imagine the rigor necessary to establish that tracks and sign left by an animal are unique to that being. Codifying such knowledge for one type of sign for even one species could take generations, and the first humans seemingly bet the entire species on the science of tracking.

There are two broad types of evidence left by an organism in the landscape, tracks and sign; the former is the print on the ground while sign tracking examines wider interrelation with the landscape. More than identifying direction of travel, size and maturity, and when the track was made, tracking is the ability to read micro expressions that detail the organism's internal state. Renowned tracker and naturalist Tom Brown Jr. explains that these pressure releases "tell if an animal's belly is full or empty, or partially filled. We can see indicators of thirst or injury. We can also see diseases, for the body also reacts to disease through compensating movements . . . Each track tells us everything about that animal: its actions, reactions, its condition, whether it is full or hungry, thirsty or tired, healthy or sick, even what it is thinking or feeling" (Brown 1999, 39). Each track contains in its contours rich descriptions of body position and movement while also detailing shockingly nuanced internal circumstances.

With experience and the context of a larger section of trackline, trackers can realize the full physiological state of the being making marks in the landscape. The depth of a track describes the fullness of belly or an injured ankle, information determining an animal's motivations and where it next moved. Granular buildup along the track's ridges register head position, indicating everything from casual interest to hint of a mate, prey, or predator, or the full intention to travel that direction. Degree of buildup on the track wall will describe the scenario, even through footwear. While persistence hunting, the Bushmen compare their own tracks to the hunted animal's—looking for tiredness indicators like stride length, kicked up dirt, and drag marks—making sure they have not lost their own bodies in the hunt, themselves approaching hyperthermia (Liebenberg 1990, 38). If the Bushmen's tracks "looked as bad as the kudu's" they would cool their heads, slowly sip water, and "then they'd walk and check their tracks again" (McDougall 2010, 237). A track is a precise description of an organism in that moment. Reading tracks is witnessing the biotic being.

Tracks describe biotic identity by registering an organism's physical state and response to stimuli. To understand or pursue that being, a tracker considers the immediate environment. Dirt is but one location the body will register in the landscape. Any tracking that happens above the track ridge is sign tracking. Moving through any space, organisms contact more than the ground. Animals rub against plants or tend to use the same hand to turn a doorknob, leaving behind hair and clothing thread. When assessing scat, Liebenberg is sign tracking. Moving above the track, sign tracking realizes the animal in its environment and "involves finding the elements of a land-scape that, when combined, make up the foundation for abundant animal life . . . a rich variety of vegetation, the availability of thick cover so that an animal feels secure that it can escape its enemies and raise its young, and finally, but not always necessarily, the presence of water. Though many ani-mals obtain their water from succulent or dew-soaked browse, the presence of flowing or standing water will create a teeming biome" (Brown 1999, 7). The act of tracking is always an examination of the surrounding ecosystem.

The presence of an animal metonymically indicates ecology, the presence of its food sources, its predators, terrain types, and climate. Understanding the needs of an organism locates that organism in its biotic context. The Kalahari Bushmen

> devised ways to run down game in every weather. In the rainy season, both the duiker antelope and the giant gemsbok, with its lancelike horns, would overheat because the wet sand splayed their hooves, forcing their legs to churn harder. The four-hundred-pound red hartebeest is comfortable in waist high grasslands, but exposed and vulnerable when the ground parches during winters. Come the full moon, antelopes are active all night and tired by daybreak; come spring, they're weakened by diarrhea from feasting on green leaves. (McDougall 2010, 239)

There is no surprise when we find that animals are easier prey in certain sea-sons, but even the phase of the moon informs the bodily ability of a species. Tracking an organism contextualizes that being in the total interrelations of its ecosystem, accessing biotic identity.

Tracks function metonymically, standing for and expressing the being who made them in response to their immediate environmental circumstance. The representation of complex ecological interrelations through tracks con-stitutes this ecometonymy, the process of tracing an ecosystem from any of its singular extensions. The physiological information registered in the track expresses that organism entirely. Tracking "bring[s] the animal back to life, as it was in that point in time when it made the track" (Brown 1999, 39). Seeing the track and sign is seeing the animal. But for timing, that track is that animal, standing in front of you. As more tracks and sign are uncovered, the organism becomes more complete, individuated and inextricable from the other organisms and abiotic features of the ecosystem. While the crossover is plentiful, and a tracker will not actually separate tracks and sign, note that tracks are intimately registering the physical animal while sign witnesses its tethers to the larger area. Tracks state that an organism must exist, and that organism states that other organisms exist, that water sources and shelter exist. Track for long enough and you will note the difference in the movements of an animal headed for water versus after drinking. You will come across a place where the being you track stopped suddenly, alerted to something, and you will go off and find the tracks of whatever caused that distraction, aging them to piece together the narrative. Eventually, you will track an animal going about its normal business and see a predator's stalking tracks overlaid on top of those you follow. Tracks are ecometonymies standing for the biotic interactions of organisms and their environs, prioritizing identities formed in interrelated nature.

Tracking functions apart from ontological constructions of self and notions seeking to elevate humans to animals. At any random, single track, only the immediate circumstances are readable, and many questions are unanswerable without significant context. Addressing why the organism is in the area or why it has a lung infection are difficult questions. This coupled with most tracking experience coming from following nonhumans, tracking creates a fascinating pocket of apolitical existence. Tracking grants access to bodily circumstance, especially as it responds to immediate environmental stimuli far more thoroughly than it witnesses political ontologies: "Not only are the slightest movements of animals and humans registered in the tracks through the pressure releases, but so are thoughts. If an animal even thinks of turning right, for instance, it will register in the track. So too, do emotions, like anger, fear apprehension, and joy, register in the track. The body reacts to thoughts and emotions in clearly defined pressure release maps. There is a body reaction clearly seen in the track" (Brown 1999, 38). While there are ways, it is far harder to read disgust with systemic injustice or various political leanings in a track, although these may describe why a track looks the way it does or appears in that spot.

The synonymity of tracks and organisms makes them indistinguishable ecometonymies. Built into tracking is a certain objectivity, a distance from concepts more abstract than organisms in ecosystems. As that is our very dilemma at the end of the Anthropocene, deploying tracking as an ecocritical framework becomes quite timely and universal. Witnessing a track accesses a brief apolitical spark of an organism in that moment relating to the immediate biotic circumstances of the environment. Recognizing the ecometonymic expression of organisms prioritizes the immediate and authentic biotic home of an organism. Ecocriticism's desire to derive and inform necessary human relations in fiction and nature is furthered, so we might identify and capitalize on moments in which we can alter the course of the Anthropocene to become an epoch where all organisms live and die well together, instead of an extinction event that closes the Holocene.

THE WIND TAKES A SIDE

Buell identifies that humans cannot speak as nature, issuing instead self-awareness of human inclusion in Aldo Leopold's "biotic community" as the closest one can come to speaking as nature (Buell 2005, 8; Leopold 1949). Following this trail through ecocriticism, Morton's ecological thought acts as a perspective, if not rigid methodology, for thinking with nature's decentralized totality, moving human thought to function as the critical interconnectivity of nature. This orbits Haraway's thoughts on staying with the trouble and the procedure for refining the thoughts by which we think through connectivity in order to live and die well together with the Earth system. The notion of speaking as nature would convert nature into human process, a metaphoric turn running counter to ecocriticism's axiomatic reengagement with real nature.

Expressing nature with human language must be the anthropocentric expression of human cultural perspective. Buell notes that we can "speak ecocentrically," not as nature but as those who culturally and ethically value nature (2005, 8). Schaberg positions the Anthropocene as a concept "to call out humans for disproportionate impact on the planet," pointing to the consistent plot of discordant relation between humans and nature in so many blockbuster films (2020, 152). Mathematics, that most objective abstraction, could even prove anthropocentric if contrasted with an extraterrestrial iteration (Clark 2018, 33). Anthropocentric mathematics might be a harder sell, but the point is taken—human efforts, even those rooted in the maximum objectivity of organic science and mathematics, are driven by human perspectives—and if mathematics can conceivably be anthropocentric, language must be less objective. Maintaining the ethos of speaking as nature with a willingness to fall short to just live and die well, the environmental humanities are concerned with how humans might think as nature to find the actions that shape the anthroposphere well. Acknowledging the unreality of gaining such primal relation to nature embodied by a feral child raised by deer and the anthropocentric murk of living as the Parhams, I would progress the shift toward thinking as nature by examining how tracking, as ecometonymic access point to nature, enables actionable confrontation with the formative aspects of the Anthropocene. Cormac McCarthy's *Blood Meridian* entangles thinking with and as nature, and tracking becomes preservative and actionable for the kid and nature to oppose the Anthropocene. Needed now as we approach reiterating the Anthropocene elsewhere in the universe, the kid entrenches himself in nature without totally evacuating anthropos or dismantling the anthroposphere.

In the penultimate confrontation of *Blood Meridian*, the kid, fully integrated with the landscape through tracking, refuses the opportunity offered by nature to kill the judge. Initiating his biotic integration, the kid rejects commerce and food extended by the judge. The kid and the expriest, Tobin, abandon the human-made well over which the judge holds court and shortly arrive at a creek surrounded by cast-off material and livestock bones. While drinking from the natural seep, the kid is fired upon by the judge, who has now acquired horses and rifles to complement him and the, so-called, imbecile companion. Opposite the judge's equipment based survival philosophy, the kid entrenches himself in the biotic identity of the natural watering hole, moving "into the creek on his belly and [he] lay drinking. . . . Then he moved out the far side and down a trampled corridor through the sands where wolves had gone to and fro" (McCarthy 1985, 300). The kid continuously maneuvers into, through, and out of the creek, mirroring wolfen action amid the tracks of these apex predators routed by westward expansion. The kid drinks at each opportunity, entrenching himself in the biome. The kid's vision is hyperattuned to the tracks and the water for the entire confrontation, and his biotic relation to the landscape manifests opposite the judge's anthropos. The only dissonance in his commitment to the creek is the stiffening of his leather clothes—a secondary over-skin purchased on credit issued to the Glanton gang for scalp hunting, a restriction that favors the judge as it was acquired by the civilizing systems he maintains.

Deeply resonating with the otherworldly scene opening *The Crossing*, the kid is double to Billy, akin and opposite. In the tracks are the actions of wolves who maintain the ecosystem by maintaining themselves. The tracks ecometonymically express wolves succeeding within the place, sufficiently eschewing the impossibility of the wolf seen in *The Crossing*. The kid becomes an organism entrenched in expressing the landscape. The judge offers friendship during the harrying exchanges while the kid bellycrawls and watches "a small caravan of ants bearing off among the arches of sheepribs. In the watching his eyes met the eyes of a small viper coiled under a flap of hide" (McCarthy 1985, 302). Unlike Billy, the kid is linked to the biotic community when doubled by the gaze of predators who are maintaining proto-iterations of human interaction with the landscape. The kid sees ants enacting the perfect, biotic version of caravanning through the desert having already succeeded at the water source where humans somehow fail. The ants then secure themselves and secure the biotic cycling of the biome by the dead livestock of failing human caravans. With an awareness and gaze preceding the kid's reciprocal recognition, the pit viper ecocritically embodies and narratively foreshadows how the kid gains predatory advantage over the judge. Entrenched in the biotic community and landscape, neither the ants nor rattlesnake alter their activity for the kid as his physical action and presence, as well as motivation, biotically respond to and secure the ecosystem. Still wolfen, the kid wipes the creek water from his mouth and continues to move against the judge.

Tobin counsels that the kid kill the horses and the imbecile, the only resources of the judge's they can affect. The expriest explains the horses are the judge's only bait and must also realize that the judge can now overtake them or travel while wounded. The horses provide the judge an additional advantage: they reveal the kid's location. They consistently alert to the kid's presence, structurally epitomized when "The kid had swung the powderflask around to his back to keep it out of the creek and he held the pistol up and waited. Upstream the horses had stopped drinking. Then they started drinking again" (McCarthy 1985, 301). These sentences, a paragraph unto themselves, are simultaneous and causal. The kid acts, and—in the supporting details that grammatically "had" already occurred—the horses cease drinking, even though they wish to continue. Unthreatened, they return to drinking. They are the judge's animals, and they reveal the kid, acting opposite the ants and rattlesnake who ecometonymically manifest the innate biotic processes of the landscape.

The horses continue to reveal the kid by watching him emerge from the water. As he cocks his pistol, "they pricked their ears and began to walk toward him across the sand" (McCarthy 1985, 301). The horses are attuned to the environment and approach the violent precursor of a drawn hammer. This is understandable, "the horse being the only animal who has adopted our ceremonies as its own" (Le Guin 1975, 277)—even war, for horses are not innately tolerant of gunfire or drawn to the mechanical sound immediately preceding. A horse must grow accustomed and learn these sounds signify a purpose and security granted the animal by humans. The kid kills these extensions of the judge, of anthropos. The judge demands compensation for his property, and the biotic kid is unmoved. He is hardly the first thing to kill livestock brought to this creek, as all desert water "was marked by the carcasses of perished animals in increasing number . . . as if the wells were ringed about by some hazard lethal to creatures" (McCarthy 1985, 308). The hazard is the biotic cycling of a teeming biome surrounding any water in a desert. The lethality of the place strikes surest those oppositely entrenched in anthropos. The lethality of wolves and pit vipers that detail the very life of the place is not visited upon the kid. The kid acts against the judge and perpetuates the creek's biotic cycling.

Preceding the horse killing, the kid curiously encounters the imbecile who had seemingly taken up with the judge. Crawling to kill the revealing horses, the kid encounters the imbecile:

> For all his caution he found the idiot watching him before he ever saw it . . .
> it was watching like a wild thing in a wood. When he looked back it was still
> watching . . . and although there was no expression to its face yet it seemed a
> creature beset with a great woe. (McCarthy 1985, 303–304)

This exchange structurally maps to the forthcoming viper scene and charts the
kid's distance from the judge as he becomes more entrenched in the biotic
wild. Ecocritically though, the so-called imbecile acts in response to the kid
maneuvering against the judge and the horses. The kid describes the imbecile among the horses as "some dim neolithic herdsman" (McCarthy 1985,
301), a far more useful categorization. The so-called imbecile is narratively
transcribed, transformed, into an historic piece in the trajectory of anthropos,
a precursor to the judge and the kid. Pivotal in this fight, a confrontation
between the judge's civilized anthropos and the kid's wilderness variant set in
a site of biotic exchange, the so-called imbecile now gazes so the reader may
recognize a version of the neolithic human extricated from the futurity of the
judge. This neolithic human, woe-stricken, watches the kid without alerting
the judge. In not attaching himself to the kid as he does the judge, he secures
the kid's survival. The act binds the neolith to the judge, an act that instead
promotes the kid's modern biotic relation.

The neolithic precursor undermines the judge, as a primal and wild human
always deconstructs anthropos. This neolithic human, an imbecile only when
transposed to the human civilization, is witnessed as "dim neolithic herds-
men" but, nonetheless, an inheritor of that same Promethean fire that becomes
the anthroposphere. The actions of such a herder directly provide the human
species with horses made unwild, one of the civilized contours contemporary
to the judge and the kid. Neolithically entrenched in biotic environs, he is
precedingly more aware than the kid, just as the viper. McCarthy also calls
"dim" the wolfen otherworld of *The Crossing*, which indicates the essence
of the wolfen and neolithic iterations of herding are equally faded in the
Anthropocene. The so-called imbecile is dim, like the biotic otherworld, as
herding is excised from biotic maintenance to constitute the Anthropocene.

Blood Meridian's postholing epilogue shows a rancher manifesting the
border of already parceled out land by penetrating the earth for a wire and
wood fence that keeps domesticated animals in respective ranches, while the
notion of property described by the fence separates other humans and their
livestock. The epilogue might as well be Parham's undescribed first year
in *The Crossing*. The wolves perform their perfect sort of herding and the
neolithic herder learns the same, both now dimmed by their modern impossibility. This dim neolith is positioned as a human inculcation of the wolfen
otherworld capable of preceding either the judge as agent of civilization or
the kid as agent of nature. Then, leashed by the judge, the dim neolith "sniffed

at the air, as if it were being used for tracking" (McCarthy 1985, 310). The judge constricts the neolithic being's relation to nature and is immediately unsuccessful at pursuing the kid. The judge seeks to domesticate the neolith as a sort of canine and presses him into tracking down the kid to snuff out opposition to the judge's anthropos—which fails, because the neolith is not evolutionarily suited for parsing the kid's scent in the landscape, from which the kid is currently indistinguishable.

The primal and wild human precisely undermines civilization by its very invocation. Derrida deconstructs Western philosophy itself in the act of needing a debase and disgusting wild human within the city for sacrificial casting out to sanctify and resecure borders (Derrida 1981). We know this is merely symbolic, and we know that it dismantles the very concept of a city as separate from the wild. Derrida shows that the very maintenance of the city depends upon the maintenance of an external wild to be manifested within.[8] Extending this, Le Guin imagines the utopic city of Omelas, dependent on another of literature's imbeciles, a child locked in a basement closet now "feeble-minded. Perhaps it was born defective, or perhaps it has become imbecile through fear, malnutrition, neglect" (Le Guin 1975, 281). Le Guin secures the city by implicating citizens in maintaining this minimum evil, and the city is metonymically undermined by inspiring self-disgust in the citizenry. Le Guin offers a reversal with the titular citizens casting themselves from the city. Walking away from Omelas, "the traveler must pass down village streets, between the houses with yellow-lit windows and on out into the darkness of the fields"—wonderfully, in a word, creating a murky and transitional space between agricultural and wild fields—"Each alone, they go west or north, towards the mountains" (Le Guin 1975, 283–284). Contained in opposing the wilds is a contortion of the wilds. Considering a spectrum of human relation to wild nature, the primal positivity of the wild child in Malouf grants Ovid transcendental vision of biotic integration that stands opposite the consumptive destruction of the judge in maintenance of the Anthropocene. The imbecile has already been doubled by the kid and the reader realizes the imbecile's own theoretical deconstruction, just as Billy saw his impossible maintenance of the wild realized by wolves. The judge's authority in the wild is always already deconstructed by his opposition to it, and the dim neolith sides with the kid's relation to the environment, as if selecting his future evolution. The only thing not deconstructed is nature itself, infinitely preceding and exceeding, only dissipating with the very fabric of the universe.

The kid, as tracker, is fully opposite the judge and indecipherable from the landscape. Extricating themselves from the creek, the kid and Tobin move into the "desert absolute and it was devoid of any feature altogether and there was nothing to mark their progress upon it. The earth fell away on every side equally in its arcature and by these limits were they circumscribed and of them they were locus" (McCarthy 1985, 307–308). The desert absolute stretches to all spatiotemporal boundaries, as though their battle is for the biotic world itself. The scene, titled "The wind takes a side" by the chapter headings, offers affirmation of the kid's tracking action as access to the biotic world. At the creek, he was ingrained in the environment by his tracking. As he and the expriest walk, the kid notices that their trackline is being taken by the wind, naturally brushed out and made identical to all other stretches of the desert absolute. With this knowledge, the pair hides from the judge and the kid makes his and our mistake.

Misunderstanding his relationship to the total environment, the kid does not kill the judge. Hidden in and as nature, the kid mistakenly uses his predatory advantage to evade the judge instead of acting as nature and killing him. He and Tobin wait among bones as "sated scavengers" (McCarthy 1985, 309) who survive on leavings despite that these are the bones of animals who have not survived westward expansion. The judge is an excellent tracker and is still confounded by this dissipation. The judge never tracked for his own biotic integration, instead seeking to dismantle nature's power over human life. He argues that "Only nature can enslave man and only when the existence of each last entity is routed out and made to stand naked before him will he be properly suzerain of the earth" (McCarthy 1985, 207). His landscape-based tether is severed by the abiotic circumstances rapidly decaying the trackline. The wind, nature, took a side.

Ecocritically, the kid accesses his immediate and potential relations to the environment by recognizing the dissolution of his tracks and gains an advantage the judge cannot penetrate. Nature has already demonstrated this exact scene for the kid when a bear suddenly rises out of the mountain forest. This unseen bear was eating carrion and, upon standing, the horses spook and the gang fires on the bear. Casually, the bear plucks one of the Delaware off his horse and plods off into the woods with the man in his mouth. For three days the gang tracks not the bear itself but the blood and sign of the Delaware carcass the bear carries, never finding either for "like some fabled storybook beast . . . the land had swallowed them up beyond all ransom or reprieve" (McCarthy 1985, 143–144). The kid can be that bear, an agent of nature consuming that which would consume the wild. Instead, he and Tobin act as scavengers sated by the corpses of livestock. Unlike for the bear though, there is no sustenance to pull from the bones among which they hide. The kid acts as buzzard instead of the bear, the predator he is positioned to be. After settlement, livestock no longer passes through these deserts to feed wild scavengers, and the predators are hunted to extinction. Such wild scavenging runs dry, and those organisms become the buzzards of *Blood Meridian*'s global south. These carrion birds are fat and sated bishops consuming a massacred congregation (McCarthy 1985, 62). Disintegrated from natural biotic cycling, the sated scavengers maintain the judge's civilizing ethos that grants bounties for scalps of first-peoples, or those that pass as indigenous (McCarthy 1985, 102–103). Scavengers disappear or become maintainers of the judge's Anthropocene.

The kid does not act biotically against the judge, the codifying remnant of the Glanton gang, as the bear does—so-named Old Ephraim (McCarthy 1985, 142) to invoke America's most prolific bear, legendary for its consumption of humans and three-toed track.[9] Despite the bear already feasting, the mere presence challenges the authority of the gang in these wild environs. The fight ensues and the bear is granted additional, fresh prey for no other reason than the gang encroached upon biotic wilds as though there they had authority. The kid does not kill the judge, incorrectly reciprocating onto the judge instead of the environment the boon the young viper gave the kid in letting him pass. The rattlesnake recognizes an apex predator hunting and entrenched in biotic action. As it would not strike a passing wolf or bear, the pit viper takes no issue with the kid. Every time a rattlesnake uses its venom in self-defense, it goes without a meal. The kid acts as the rattlesnake did but fails to embody, to think as, the viper which will forego hunting to kill something threatening its life. The kid's current circumstance is that he and the landscape are under threat by the very presence of the judge, who is within striking distance.

Able to leave the desert, the judge ushers in the Anthropocene, that epoch constructed by things cleaving humans and nature, contranymically separating and binding—capital, globalization, crafted hyperobjects, and the rest of the anthroposphere used to simulate independence from nature but integrate through the biotic cycling of extinction. The kid, in some delirium, sees truly the judge overseeing

> an artisan . . . a worker in metal . . . a coldforger who worked with hammer and die, perhaps under some indictment and an exile from men's fires, hammering out like his own conjectural destiny all through the night of his becoming some coinage for a dawn that would not be. It is this false moneyer with his gravers and burins who seeks favor with the judge and he is at contriving from cold slag brute in the crucible a face that will pass, an image that will render this residual specie current in the markets where men barter. Of this is the judge judge . . . (McCarthy 1985, 322–323)

He is the judge of doubles who are alike enough their true and preceding other to act as thoroughly, but in systemic deterioration instead of maintenance. This artisan, without even the Promethean fire that doubles the internal fire of the wolfen otherworld, crafts second-level doubles. In the Anthropocene,

we are doubled forgeries of ourselves as we attempt to live as creatures integrated with an ecosystem that no longer exists.

The judge ushers in a world of every kind of capitalist extension and per-
petuator of war, which is a doubled forgery of biotic violence. He holds court
over these facsimiles in a bar where a dancing bear is shot for no reason,
or the reason the Glanton gang shot at Old Ephraim: humans shoot bears.
Random men wander to find from whom they might purchase the pelt. The
dimness manifested by *Blood Meridian*'s neolithic and wild being and his
own personal woe is ours in the Anthropocene. We gaze at ourselves and
recognize the forgeries of wildness we are, for we crafted a biotic system
that disintegrated us within it and have created an epoch our biotic selves
do not integrate with outside of necrotic decay. We know the kid, a human
agent of the wild, fails to kill the judge because he has already failed. The kid
only acts biotically against the judge to survive. Killing the judge would be
an apolitical act benefiting our biotic home. The kid cannot shoot. He would
alter history, and we would recognize our actions soon enough. We and he
did not. We are in the Anthropocene. In the final sliver of violence we are
not allowed to see, the judge consumes the kid. We knew this would happen
because *Blood Meridian* describes the formation of the United States. Look
around. The judge has authored this.

COSMIC OTHERWORLDS

We are as the kid. Huddled here at the meridian of the late-stage Anthropocene,
on the cusp of committing to the future form of human ecological control in
the Anthropocene and determining how we will relate to the natural world
in and beyond this epoch. Humans fast approach biotic interaction with the
universe beyond our biotic home, the last nonhuman wilderness—that is, the
last environment that does not contact the anthroposphere. In the late-stage
Anthropocene, humans finalize their relation to Earth and expand beyond
its limits. *Blood Meridian* describes how we handle ecosystems. We left the
wilds for the judge.

Residing in the kid's inaction is our consent to prioritize human bodies
over the biotic system that evolved them, ushering in the epochal ecology
that doubled us with ourselves. We made ourselves forgeries of the beings
that maintained the ecosystem by maintaining themselves and their kin. We
now maintain that categorical forgery of the nonbiotic human. We left alive
a man, a philosophy that promotes the notion that "whatever in creation
exists without my knowledge exists without my consent" (McCarthy 1985,
207). In this philosophy and motive, the judge copies into his book singular
images from a grand story told on stone before scratching out a single piece

of this petroglyphic indigenous (hi)story, contorting the original meaning by filtering it through his intent (McCarthy 1985, 180); he precisely sketches artifacts that he then crumples and destroys so that the image he forges is the only extant version (McCarthy 1985, 146); and he kills and stuffs birds in an act reminiscent of preparations for a gluttonous feast that never comes for it would at least reference some biotic and seasonal relation that the judge acts against, instead manifesting a non-predatory, nonbiotic, form of consumption because he cannot currently conjure the twisted zoo to house all birds to extinguish the flight offending his suzerainty (McCarthy 1985, 208). Without illusion, the judge acts for anthropos.

In *Blood Meridian*, there is a meteorite that has "wandered for what millenia from what unreckonable corner of the universe" (McCarthy1985, 251) to enter the atmosphere and collide with the Earth's surface. There it remains. This slag is used as an anvil for that Promethean forging and reforging of civilization. The meteorite expresses a universal ecosystem that constitutes a good place for humans to gather but without shifting the contours of the landscape itself. This very moment, describes how humans might interact with the natural history of the planet as it is ecologically contoured with the entire spatiotemporal expression of the universe, granting boons that remain themselves while maintaining humanity. The existence of this meteorite supposes that we might travel in such a manner, arriving elsewhere to become indecipherable from that landscape while referencing earthly, biotic identity and the universal ecosystem. I hope our fascination with earthlike exoplanets speaks to such an ethos.

We are, however, properly staged to continue as the judge would have us. The judge hauls that meteorite above his head and hurls it a distance beyond all gamblers' reckoning, winning the random wager and gathering to himself alone the literal and figurative human capital of such a feat. The judge does not tolerate the universe arranging itself without his input, without his own perpetuation. He hurls that meteor randomly, doubling whatever universal system set it on the path to this planet, winning an expansive collection of currency that he shares with no one. This is how we might interact with this last wilderness—impacting the natural contours of the universal ecosystem for personal capital, regarding the universe's wilds as we have those of our biotic home. We have created the military branch that secures and enables, however vaguely, human interaction in space (U.S. Space Force) and legal policy for asteroid mining has been instituted (*U.S. Commercial Space Launch Competitiveness Act*). My biggest fear is that the judge is inextricable from humanity and defeats our ability to balance the Anthropocene.

I wonder how we might do it well. How might we change a forceful and gargantuan, a planetary, trajectory? What comes next is written by an SF nerd who, in his heart, thinks the whole thing is awesome—Space Force, asteroid mining, terraforming, living *Star Wars*; damn, that is cool. This feeling is at odds with the naturalist philosophy circulated through my tracking. I seek a process by which assessing the apolitical being in its biotic home realizes and informs a biotic ethos for interactions elsewhere. In the universal ecosystem, our biotic identity is unprecedentedly displayed and challenged in ways we are only beginning to realize. I wonder how we might reveal and remember just how entrenched we are in the universal ecosystem. The combined fascination with organic Earth systems in relation to the universe drove me toward the otherworld of *The Crossing* that begins this chapter. The otherworld is constructed as much by moonlight reflecting off the snow as by wolves. The Earth is illuminated and strangely becomes moon-like. The space is witnessed by tracking the biotic metonyms and tucking one's hands into coat sleeve technology. Better yet, we can integrate with the universal ecosystem by persistence hunting an animal exhausted by the light of a full moon keeping it awake all night. The difference is merely if humans act in maintenance of self or biotic self.

Tracking benefits the ecological thought's infinite focus and the notion of living and dying well, especially as humans will imminently live off-world. These are correct and necessary perspectives imperative to informing the predictive guidance of our temporal proximity to an addendum or alteration in the biotic notion of home. Perhaps all that is needed is to shift our focus as these perspectives describe. The problems of scaling only manifest if we invoke an idea that does not actually scale. There are plenty that do, including harmful anthropocentric frameworks. (We need adamant destruction of those notions.)

On the moon sits both iterations: we left engineering marvels and a flag, and we left tracks. I would trade almost anything to read the first human track on the moon. Those prints are largely unchanged, still detailing the very actions of human bodies in a novel environment. There is no wind to wipe them out. They have probably decayed a bit at the ridges and crests as the moon's gravity attracts loose grains, although not nearly as much as the other things we left behind. In a pleasant reversal, the lack of atmosphere that so thoroughly preserves the tracks subjects everything else on Earth's moon to destructive temperatures and solar radiation (NASA 2017c; Spudis 2011). The tracks relaying immediate biotic circumstance persist in the moon's environmental system while the things we brought from Earth do not signify in the same manner.

NOTES

1. Seeing a wolf pack take down an elk is well worth the price of admission to Yellowstone and a ranch having that balance of cattle and land so they might roam comfortably is something to be desired.

2. Since we are speaking of wolves, I would note that "consuming" here functions metaphorically rather than biotically.

3. There are perhaps only two differences between humans and animals that I will entertain: this need of ethical philosophy to live and die well together, and the sheer potency of impact on the environment becoming the ability to control its planetary expression.

4. In checking that I was the first to use this term, I found only Jacques Derrida's use of "ecometonymy" in his reflection on his friendship with Helene Cixous. Derrida casually forms the word while considering an "absolute economy of speed" (2006, 73). Derrida's "eco-" is not contextually the same as my invocation, although—as must be the case in all things Derrida—the difference slips a bit. The root "oikos" travels forward to economy and ecology as a system of relations. Derrida engages the metonymic speed by which words mean, moving between the words and their metonymic expression of body in heartbeats. Delightfully, Derrida embraces the metonymic associations of "speed" to touch upon bodily movement and spacetime—critical concepts to tracking and the universal ecosystem in *The End of the Anthropocene*. Although Derrida arrives at the body more by metonymy than eco-, the fundamental differences between Derrida's and my own use of the term seem also to trace a fundamental sameness. This lexical exchange resonates deeply with tracking as a tangible identification of signs that is akin to written language, which is Derrida's target in the book. I think of the tracker Famous Shoes in Larry McMurtry's *Streets of Laredo*, who is a master tracker who wishes to learn to read and constantly refers to letters as tracks.

5. Camping, hiking, and the whole of ecotourism can be wondrous experiences but, sadly, rarely exist without the contours of civilization's destructive environmental aspects. Even the *Fourth National Climate Assessment* (2018) couches many of its arguments in the economic bottom-line—a proven rhetorical strategy, but one that this author dreads signals a "last chance" to convince the public. I have little faith in the fundamental methodology of the capitalist cityscape to preserve nature and suspect that capitalism, if not economics in the twenty-first century, must taint the environment, both actively harming and with policies preserving the capitalist tenets opposite nature. If offered, would we sacrifice any civilizing feature for the wild? I think any answer must first remember the anecdote in Timothy Morton's *The Ecological Thought* in which a wind farm is never built because it will ruin the view (Morton 2010, 9).

6. This might then be recirculated through discourses of urban ecosystems to produce a superior ecological ethos that guides solutions to problems of environmental destruction and the body—human or animal; the feral Child makes the distinction transient and irrelevant—as the victim of urbanization's slow violence.

7. Frighteningly, "our death" may no longer be an individual demarcation, and wholly interchangeable with "our extinction." Fingers crossed, though.

8. Similar Yellowstone's inclusion in national identity, wild space remains firmly external civilization. The park is external to the city itself but maintains it by travel, tourism, and notions of a vacation in the wilds as just what one needs to recharge a civilized self.

9. Utah State University's library maintains a digital exhibit and collection of legendary grizzly, Old Ephraim. You can examine a 3D scan of Old Ephraim's skull, hear recordings of the legend, examine primary source material, and find directions to the bear's grave. (https://digital.lib.usu.edu/digital/collection/Ephraim)

Chapter 3

A Body in the Universal Ecosystem

We are the only terrans to live beyond Earth.

On November 2, 2000, the Expedition 1 crew opened the hatch to the International Space Station, becoming the first humans to call it home. Since then, our species has maintained a permanent space presence on the International Space Station becoming the only Earth species to exist outside the biotic environment of our home planet. Very soon that same clock will start for the moon and for Mars.

In 2018, NASA announced the exploratory venture *Moon to Mars*, a program establishing our lunar return with the first woman in 2024 followed by permanent installations on the surface and in lunar orbit (NASA 2020b). From there, we leverage the lunar station for Martian exploration in the 2030s, followed by our Martian presence and additional deep space exploration (NASA 2020c, 2020a). We are going to live on another planet. This trajectory clashes with most of the efforts in the environmental humanities. The problems of this planet are potentially ignored, this time in favor of invoking new problems or perhaps just bringing the old ones with us to some Planet B. There are issues of misappropriated capital in numerous discourses, and the underlying distrust of capitalism in our field certainly raises an eyebrow at NASA's (necessary) private industry partners for interplanetary ventures. And yet, we are going.

Right now, we are a species simultaneously engaged in destructive environmental practices and preparations for permanent human presence on the moon and Mars. These two actions, if not already, will accentuate and alter what it means to be human. I see no lasting distinction between human philosophy of environmental interaction on Earth and elsewhere. I am hopeful that the reciprocating progression of Earth and space technology will conjure renewable and sustainable uses of the resources required for space exploration. Perhaps we will even find ways in which the use of those resources ensures biotic communities. For instance, solar power may be critical to our efforts on the moon, and Mars missions are likely to utilize nuclear fission (NASA 2020d, 2020a). Perhaps we will even travel by solar sails, capturing photons as though they were an earthly wind (The Planetary Society 2021).

Of course, there is the other thing. We can carelessly consume the finite resources within our reach, grabbing metallic asteroids and icy comets to, judge-like, control where they are and to what end they contribute. Our civilization can anthropocentrically disrupt ecosystems to expand beyond our biotic home. We have done it before, and terrifying extensions have already been imagined. Schaberg gestures at pop-culture iterations—the dystopian industrial planet, Corellia, in *Solo: A Star Wars Story* (2018), the apocalyptically polluted and evacuated Earth of *Wall-E* (2008), and the uninhabitable extraplanetary environs of *Gravity* (2013) and *The Martian* (2015)—and identifies "our disproportionate impact on the planet [is] the substratum of all these films, whether they know it or not" (Schaberg 2020, 152). Similarly, in an article on *Star Wars* and ecology, I briefly touch on the representation of weaponizing nonrenewable solar power (Gormley 2019, 41). The image of a gutted planet technologically reconstituted as a space station that consumes entire suns for galactic dominance is frighteningly allegorical (Abrams 2015). Fortunately, my estimates put our ability to construct such a star killer a few years off. These texts circulate the ecological tensions of the Anthropocene. They are either prophetic or inculcate a proper ecoethic for the end of the Anthropocene.

We are still on the cusp of the late-stage Anthropocene, of figuring out how we will ecologically comport ourselves on this planet and others. I recall Yoda's lesson for Luke at the cave on Dagobah. On a planet whose singular biome is a swamp rife with biotic cycling, Luke stands in front of a cave with unsettling potency. The cave itself is far more natural than we traditionally recognize. Its darker, but no less natural, tones are quickly overwhelmed by the imperial applications of the Force by Vader and Palpatine.[1] However, the snakes and lizards of the dark side cave do not take up their traditional symbolism and are casual beings, no more antagonistic than the wild swamp. Earlier, a snake is simply moved out of a bowl and onto the table in Yoda's hut so Luke can get himself some stew, displaying Yoda's biotic integration with the swamp—"Mudhole? Slimy? My home this is" (Lucas 1980). Luke is the source of the anxiety, unwelcome, and fear manifesting in the environment. Yoda tells Luke what is in the cave: "Only what you take with you" (Lucas 1980). Luke straps on his lightsaber and blaster—notable tools of imperial power and indicative of technologically induced distance from wild nature—and promptly finds fear and failure.

The conflict is manifested by the individual within the cave. The environment reacts to his disintegration.

The lesson is reissued for the Anthropocene in Nolan's *Interstellar* (2014), a story of humans resisting extinction on a failed Earth system by relocating to a distant but likely habitable exoplanet. Hopeful, Dr. Brand ruminates on humanity's ecoethical precipice and the need for integration with a planetary system. She parses the incredible human capacity for "evil" destruction as distinctly other than the integrative, natural violence of, say, an apex predator ripping an animal to shreds. "Just what we take with us, then," echoes Coop (Nolan 2015). And here we sit, trying to stabilize Earth and permanently live on Mars. Substratum indeed. The late-stage Anthropocene decides the eco-ethic with which we enter the universal ecosystem. Which perspective, which praxis for living on a planet do we develop, do we bring with us? How we handle our biotic home in the late-stage Anthropocene is how we will handle extraplanetary environs.

As ecometonymic expressions of our biotic home, every aspect of our species has been conjured in reciprocating relation with Earth. Expanding beyond our biotic home into the universal ecosystem quickly accentuates the biological needs met by our blue planet. Attempting to wrap the gargantuan, amoebic, and undetermined human futures invoked in climate crisis and interplanetary expansion, I adopt what science fiction has already decided: Earthlings are terrans. As ecometonymies dependent on biotic and abiotic interactions, I extend this nomenclature beyond humans to all earthly organisms. Anywhere, our life demands other terran organisms. Thus, my opening statement needs amending. Humans and the organisms comprising our microbiome are the only terrans to have a permanent extraplanetary presence.[2] By these microbes we digest food, produce vitamins, and maintain our immune system (Gopal and Gupta 2016; Biesalski 2016; Young 2017). Even imagining the ISS with only nonpathogenic microbes, those actively maintaining or just not problematizing healthy bodily function, there are still pounds of microorganisms per person that human bodies do not function without. This microbiome travels with us, surprisingly increasing in diversity in the isolation of the ISS (Gilder 2020). The organisms constituting the microbiome are critical kin paradigmatically shaping the biotic notion of the human body.

Obvious in space, human bodies are jarring ecometonymies implying necessary resources largely unconsidered on Earth. This extends beyond the only water and food for maintaining spacefaring terrans is brought from the planet or recycled on the spacecraft. Excepting those who spend significant time under water or at altitude, humans rarely consider where their next breath of air comes from. While air quality is of decided import in the Anthropocene, we do not sit in our vehicles breathing only the air we brought with us or need oxygenators installed in our homes to create the molecular balance of breathable air. Most dramatic and least considered, our biotic home provides the constant gravitational baseline in which we developed all anatomical features. Astronauts return with decreased bone density and atrophied musculature, a condition normally prevented by just living in the 1G of our biotic home, even with exercise regimens designed to stave off these effects (NASA 2021). Gravity is an abiotic contour of our home planet largely unconsidered until humans leave and return to Earth. Entering the universal ecosystem, we find gravity is a fundamental, natural contour linking ecosystems to the very fabric of reality. (This is why I drastically jump in scale from a planetary to universal ecosystem.) Who knows what else will appear as we continue to examine how the human body interrelates with the environs beyond Earth.

By this, ecocriticism and the environmental humanities extend to the universal ecosystem. The universe is pretty big, so let us move just one planet over to consider how the biotic human body integrates with the Martian ecosystem. Continuing to think of first-wave ecocriticism as primarily examining personal relations to nature while second-wave ecocriticism focuses on an invocation of scientific frameworks for nature, ecocritically reading the human body on Mars poses a challenge for ecocriticism. We have no nature narratives because we have never been, and science fiction traditionally reflects and refracts terran circumstance on Earth. Ecocriticism needs a novel technique for assessing the biotic human body in forthcoming environs.

RECIPROCATING ECOCRITICAL METHOD

Ecocriticism supports by analyses of narrative and poetry the critical efforts of the environmental humanities as it navigates current human relation to planetary and local environments while tracing the Anthropocene's impact on terran communities. These efforts refine a desperately needed ecoethic for surviving the Anthropocene. The environmental humanities is an interdisciplinary approach to assessing the current interactions of nature and human culture, and ecocriticism is its literature department. Ecocriticism lives presently, a wonderful state capturing as an ethos the biotic relation with which the field is concerned. In the Anthropocene, the present for the environmental humanities is strange. Success is opposite extinction. The work of the humanities has always been to interpret the throughlines of culture. Assessing how we respond to the environment and ethical mediation of our impact falls to the environmental humanities.

Terrans are in contact with novel environs of the universal ecosystem and will shortly expand and intensify these interactions. The ecological present now includes extraplanetary ventures that stymy ecocriticism by generally restricting access to the imminent futuristic events. Ecocriticism lacks extraplanetary nature narratives while fictional representations have greatest utility as commentaries on ecological dissonances in the Anthropocene. First-wave ecocriticism must make do with the brief lunar landing and narratives of the International Space Station, which largely focus on our scientific relation to space and the awe of being off-planet. These are engaging but not overly representative of living biotically integrated with the surrounding environs, instead demonstrating the difficulties of simulating earthly circumstance. Second-wave ecocriticism experiences a boon in the rapid advancement of general scientific literacy, allowing a deeper interaction with SF narratives by both writers and audiences. Present-minded and in the SF tradition, ecocriticism readily draws out the genre's philosophical and material warnings and potential responses to the ecological problems of the Anthropocene. Living in the present, ecocriticism is too often reactive. Ecocriticism is invaluable, and ecocriticism is hamstrung. Instead of only assessing the impact of humans on the environment, ecocriticism can predict and impact the contours of human biotic identity in and after the Anthropocene.

Imbuing first- and second-wave ecocriticism with reciprocating motion manifests a new ecocritical method capable of predictive claims about humanity's biotic and cultural identity as the species spreads beyond Earth. To produce this reciprocating action, first-wave ecocriticism must assess fictional texts that act like nature narratives, describing humans entrenched in the environs of other planets. SF has plenty of these. Axiomatically concerned with engaging actual nature, ecocriticism refines the list of texts to those placing a premium on ecorealism. Realistic depictions of extraplanetary nature are limited by having nearly no experience living anywhere but Earth and our knowledge of the universe being largely theoretical. First-wave personal narratives can stretch to include particular fictions, but only the slimmest segment of the planetary survival subgenre of SF. These stories focus on hard science fiction survival narratives on real planets whose ecology is well understood. Otherwise, the text returns to the arena of second-wave ecocriticism as something reacting to our current space in the Anthropocene. Ecocriticism's predictive ability is generated by human engagement with tangible and realistic environs.[3]

Gathering these texts requires extending second-wave ecocriticism's concern with scientific literacy beyond the earthly environmental sciences and into the universal ecosystem, where planetary geology, astrobiology, astro- and theoretical physics describe the landscapes. Extending second-wave ecocriticism's concern with scientific literacy maintains the predictive ability of this method. Always attuned to ecocriticism's interdisciplinary responsibilities, the reciprocating ecocritical method must prevent the slippage of scientific terms into metaphoric approximations that would negate the authority of its predictive claims. In this method, second-wave ecocriticism generates the human relation to the real world while first-wave ecocriticism addresses fictional representations of those interactions.

First-wave ecocriticism still assesses nature narratives but draws to this category some of the hard science fiction texts formerly managed through second-wave analytical processes—where it reflects human-environment relations to Earth in the Anthropocene instead of evidencing human relation to another planet. Second-wave ecocriticism grows its scientific literacy to parse authentic simulations of biotic human relations to other planets, offering texts to first-wave ecocriticism that will not exist for at least a decade while constantly drawing into second-wave ecocriticism the scientific knowledge necessary to describe and affect interplanetary humanity. This reciprocating ecocritical method evolves the standard model of ecocriticism as composed of two mostly independent perspectives for assessing human-nature relations. Setting the waves of ecocriticism to orbit each other in a process of recursive reference for informing each, ecocriticism predicts the contours of human interaction with other planets and impacts the form such relations take. Additionally, the method can be recirculated to support the current ecoethical efforts of the environmental humanities.

By this method, I would set first- and second-wave ecocriticism spinning about each other, each drawing materials from and offering direction to the other. I like to think of ecocriticism as a binary star, a two-star system in which the stars spiral around each other constituting between them a center of gravity for the solar system, which becomes the environmental humanities in this musing. The image relays many things. The environmental humanities forms around ecocriticism while acting as a defining container. Binary stars share material, and one eventually consumes the other. Luckily, this does not alter ecocriticism as it does not matter if the waves become just one thing. Plus, we can just keep them separate if we want. My dramatic sensibilities and general fascination with binary stars may not be so helpful for you imagining the reciprocating method. If so, imagine an ink drawing of two curved arrows cyclically pointing to the others' initial point.[4] I like how this cycling references and refracts the deconstructive action. Here it promotes and reinforces ecocriticism rather than dismantling it—a binary not in opposition. Perhaps one last embodiment of the reciprocating ecocritical method will help.

MARS is a 2016–2017 *National Geographic* series that, in classic documentary fashion, bounces between the narrative arc of the topic and the experts on the subject—extraplanetary travel and colonizing Mars. Responding to our lack of Martian narratives, *MARS* develops a fictional narrative describing the first humans launched to colonize the red planet in 2033. The series delightfully jumps to the present day to include analyses and context from astronauts, JPL engineers, authors, politicians, and private sector personalities. In one such interview, Andy Weir says, "We need to go to Mars because it protects us from extinction . . . once humans are on two different planets the odds of extinction drop to nearly zero" (MARS 2016). The episode follows up with Elon Musk describing SpaceX's use of NASA Launch Pad 39A, the "hallowed ground" from which Apollo 11 launched, putting humans onto the moon for the first time. Episode one intersperses this commentary with footage of the astronauts boarding Apollo 11. Musk echoes Weir's concern of being a non-spacefaring civilization stuck on a single planet for an extinction event. The fictional astronauts echo these ideas and then contrast with hope, wonder, and enthusiasm life on Mars. The series' structure is inspired by the blend of storytelling that puts people on Mars and the technology that sustains them in Stephen L. Petranek's TedTalk turned book, *How We'll Live on Mars*.[5]

MARS actively deploys a fictional nature narrative unfolding on Mars as moderated by and moderating the scientific knowledge used to sustain terran bodies on the red planet.[6] Here entwined are human fears in the Anthropocene, biotic entrance and interaction with another planet, becoming a spacefaring civilization, human need and desire to live on Mars, what it means to be human when we get there, and what we are and can be once we do. Instead of refracting the present, it builds the future from it. Ecocritical analysis uses this same reciprocation of personal narrative and scientific prediction to examine terrans in and after the Anthropocene.

CURATING A CROSS SECTION OF SF

Humanity is about to make an unprecedented foray into the solar system, and we need to do it well. The reciprocating ecocritical method describes a nuanced cross section of SF needed to posit the future interactions of the human body in the otherworldly environs of the universal ecosystem. Texts need to act as first-wave nature narratives while being validated by the extended scientific literacy of second-wave ecocriticism. Lacking interplanetary travel memoirs, we need a method for curating a cross section of SF rooted in texts that abide the physical universe. Twenty-first-century hard science fiction is the obvious starting category. Hard science fiction derives much of its authority from its rigorous scientific plausibility. Of course, science fact is a shifting paradigmatic aspect of culture. This and the next chapter limit the predictive applications of the reciprocating ecocritical method to twenty-first-century texts.[7] Additionally, local terrain, geography, geology, ecology, and climate become dramatically speculative the further out from Earth we go, challenging the scientific accuracy necessary for a text to act as a nature narrative. We are headed to the moon and Mars, and so the texts read here stay in that realm.

Accessing the breadth and opportunity of science fiction in *Staying with the Trouble*, Donna Haraway simultaneously abbreviates and reconstitutes the genre as the near infinite SF. SF "stands for" in a fabulously productive play of meaning. Haraway uses SF to mean "science fiction, string figures, speculative fabulation, speculative feminism, science fact, so far" (Haraway 2016, 2)—adding to it as she finds new visions and versions, leaving us to chuckle and ponder the multiplicity of "so far" as it playfully implicates the reader. The play of meaning Haraway conjures is perhaps the best example of productively utilizing the deconstructive pile of meaning as it gathers too much to itself. SF attractively enables work instead of destabilizing meaning for it is first, but maybe not primarily, a genre, and anything meant must be authentically entwined. SF always means all its things. That which we examine grants access to everything else—as witnessing mice droppings implies an owl or snake or examining grazing herd fauna is to know an apex predator.[8] SF becomes "a method of tracing, of following a thread in the dark" (Haraway 2016, 3). The abbreviation capitalizes on its origins while rousing within the containing genre every creative enterprise of fiction. And isn't that what we need? We are a species simultaneously engaged in destructive environmental practices and preparations for missions to Mars. These are swampy times, and it's hard to move in swamps.

The End of the Anthropocene, however, creates a path through the muck and murk by making use of two particular SF threads to identify texts suiting the reciprocating ecocritical method—specifically enmeshing two of Haraway's SF tentacles: science fiction and science fact.[9] Applying Parini's description of ecocriticism as "a re-engagement with realism, with the actual universe of rocks, trees and rivers that lies behind the wilderness of signs" (Parini 1995) curates a cross section of SF that prioritizes modern texts placing a premium on scientific accuracy to maintain a hypothetical viability while imagining human futures. Call it science fact fiction. I prioritize fact only for a moment because they must be scientific facts for this method, but the categories are reciprocating, interchangeable, and inseparable. Like and as SF, I always mean both fact and fiction. I find this subgenre of SF useful because it parallels realism's relation to fiction. Realist texts are not memoirs. They use fiction beholden to and describing the nuance of real human experience. Likewise, science fact fiction stories are beholden to the scientific realities of our universe to ruminate on human experience.

Michio Kaku's delineations of science fiction technologies are particularly helpful when maneuvering this cross section of SF. Kaku establishes three classes of technological impossibilities. Class I impossibilities are those "that are impossible today but that do not violate the known laws of physics" while Class II impossibilities "sit at the very edge of our understanding of the physical world" (Kaku xvii). Kaku categorizes, with some accommodating definitions, things like force fields and teleportation as Class I impossibilities, while placing time travel and wormholes in Class II. Class III impossibilities "violate the known laws of physics. If they do turn out to be possible, they would represent a fundamental shift in our understanding of physics" (Kaku xvii)—think perpetual motion machines. Science fact fiction loves Class I and II impossibilities. Abiding physical law, Class I and II technologies evidence the contours of the universal ecosystem, especially as this book explores an interplanetary humanity. Conversely, Class III impossibilities are of no concern as they do not enable a "re-engagement with the actual universe" (Parini 1995).

From these guidelines, the remainder of this chapter performs close readings of Andy Weir's *The Martian* as a nature narrative of the red planet. In the novel, engineer and botanist Mark Watney is left on Mars after being presumed dead by his Ares 3 companions. *The Martian* is a classic iteration of the stranded on an alien planet trope, gathering to itself a dramatic immediacy as NASA prepares the very mission types Weir diligently imagines. In classic ecocriticism, we read Watney's difficulty surviving on Mars as an allegorical expression of our opposition to earthly environs in the Anthropocene. Noting Watney's use of technology to resist the natural Martian environment, he implements a perfect environment for his crops—technically colonizing Mars (Weir 2014, 147)—and a microcosmic installation of the Anthropocene. Such a reading is exactly why the environmental humanities and science fiction are so productively entangled. We are dangerously capable of doing to Mars what we have done to Earth. This is a potent path to travel, but it utilizes ecocriticism to comment on the terran present. Instead, I read the Martian environs as the actual landscape of Mars so the reciprocating ecocritical action can interrogate human biotic future in the universal ecosystem.

Read with the reciprocating ecocritical method, *The Martian* becomes a nature narrative describing the challenges of humanity's first extended maintenance of an interplanetary human body. Structurally, *The Martian* is already adapted for first-wave ecocriticism as the narrative is largely Watney's first-person journal detailing his interactions with Mars as it challenges his technology and his terran anatomy. Watney faces his body's inability to survive the baseline atmospheric conditions of Mars and its more extreme climatic activity. For second-wave ecocriticism, Weir's well-researched approach to Mars missions imbues the novel with a realistic, hypothetical status. While not how NASA plans to go about it, there is little complaint by the scientific community that Weir's imagined technology could get Watney to and keep him alive on Mars. Watney's farming methods are more highly scrutinized, circulating in pop culture and scientific circles as any iconic hard SF scene does, investigated and reimagined as we learn more about Martian soil composition and ecology (Wamelink et al. 2014; Shamsian 2014; NASA 2017a;

Wageningen Environmental Research Institute 2017, 2020; Romero 2020). *The Martian*, when read with the reciprocating ecocritical method, reveals ecocriticism's ability to identify and impact the next human paradigm and the epoch succeeding the Anthropocene.

THE BODY ON ANOTHER PLANET

In the late-stage Anthropocene, we establish a consistent presence on Mars. Evolved by and in Earth's atmosphere, entering an alien planetary ecosystem is enabled by our technological ability to simulate Earth in microcosm, dispensed individually as various spacesuits and communally as habitats and vehicles. Terrans spreading into the universal ecosystem necessitates technological simulations of earthly environs. If we ever do, we are a long way from visiting a planet similar enough to Earth that we might expose our skin to the atmosphere and breathe unaided. Traveling through space to such places always demands some sort of ship and suit capable of simulating our biotic home's atmospheric pressure and air content. Traveling to Mars and living on its surface demands spaceships, habitats, planetary vehicles, and suits that approximate our home planet.

Regaining consciousness alone on Mars, Watney's first crisis is that his suit is breached by the communications antenna that slams into him as the crew evacuates the planet during a storm that would topple the only rocket capable of getting them off Mars. His suit and body punctured "the copious amount of blood from [Watney's] wound trickled down toward the hole. As the blood reached the site of the breach, the water in it quickly evaporated from the airflow and low pressure, leaving a gunky residue behind. More blood came in behind it and was also reduced to gunk" (Weir 2014, 5). Landing facedown torqued the antenna against the hole in the suit. The coagulate gunk and position combined to form a seal good enough that the suit could manage the pressure drop. Watney explains that his suit's life support is most limited by its ability to filter CO_2. He was unconscious long enough that his filters were spent, so the suit went into emergency mode, venting air to Mars and backfilling with Nitrogen which also runs out. The only thing left to pump into the suit to maintain atmospheric conditions was O_2, so the suit does. Watney notes the irony of a terran dying on Mars from too much oxygen in his punctured spacesuit. Earthly atmospheric conditions are incredibly fragile elsewhere in the universal ecosystem, even considering how disrupted the atmosphere is in the Anthropocene.

In the communal space of the Hab, short for habitat, CO_2 is managed for the crew by the oxygenator which breaks up the molecule and reclaims the oxygen. Too large for a spacesuit, the oxygenator helps maintain atmospheric conditions in the part lab, part residence that is the largest simulated Earth space on Mars. Technologically simulating earthly environs conjures an odd intermediary space of earthly and Martian planetary conditions. What can be simulated shifts based on where terrans are in the universal ecosystem. Humans must have the pressure and chemical composition of Earth's atmosphere while likely just enduring some atrophying effects of living in roughly a third of Earth's gravity. Suddenly, gravity is a resource with fluctuating availability across the universe. On Mars, other resources become abundant. Watney describes that because of "a neat set of chemical reactions with the Martian atmosphere, for every kilogram of hydrogen you bring to Mars, you can make thirteen kilograms of fuel" (Weir 2014, 3). Also, solar power is critical to the Mars missions, powering everything in the Hab, including the lights that simulate Earth sunlight—the same sun lights Mars but suits disrupt any absorption while the Hab simulates the atmospheric disruption of sunlight found on Earth. Being terran in the universal ecosystem is inextricably technological. This is true on Earth too, but elsewhere in the universal ecosystem our cyborg parts approximate planetary characteristics necessitated by our biotic bodies.[10]

Living and dying in the mediated space created by simulating Earth on Mars, Watney ecometonymically describes both his biotic home and his immediate environment. The suit he wears outside the Hab is maintaining pressure on his body ranging somewhere between the minimum survivable pressure of the human body and average Earth pressure, all within the low but existing atmospheric pressure of Mars. The suit also protects against solar and cosmic radiation that is mitigated by our atmosphere. Solar panels capture this sunlight and the crew powers the entire camp with this energy, including lights simulating Earth's sun. In Martian gravity, the fastest way for humans to travel on foot is skipping instead of the running we are biomechanically designed to perform (Weir 2014, 167; McDougall 2009, 220–221). Ecometonymically, the biotic body is detailing the environment of Earth and Mars. There is no abiotic organism, only novel environments included in biotic action.

Thinking ecometonymically entangles the biotic home and the biotic location to make a third, mediated site that manifests the ecoethos that moves terrans into the late-stage Anthropocene and universal ecosystem. Just as on Earth, human interaction on Mars can swing in either direction. Humans can consumptively engage or live in concert with the environment. In the first method, these Martian missions utilize a highly unstable isotope of plutonium to generate power for the vehicle that will get the crew off Mars. Once the power is generated though, they just bury the box of plutonium far enough from the site. (The action reeks of slow violence.) Alternatively, the Hab does not need a cooling system as there is no shortage of heat loss to the Martian air.

Recalling the image of Watney dying on the Martian surface, Watney's body ecometonymically details, through predictable chemical reaction, the dissonance of earthly organisms in the Martian atmosphere. Watney's blood rapidly changes state as the water fundamental to life on Earth instantly evaporates. Oddly, exposure to the Martian atmosphere while lying face down in the dust, the biotic interaction between Mars and Watney saves his life. Human biotic identity is constructed by relation to the ecosystem the species evolved with and in as well as the ecosystem the body is within. The Anthropocene doubles us on our own planet, an apparent motif in and beyond the late-stage Anthropocene. The universal ecosystem similarly doubles us by describing terran physiology formed to suit elsewhere. Terrans are doubled, biotically unbalanced in the total universal ecosystem.

MARS GAZING

Watney's circumstance is continuously framed as Mars trying to kill him, the planet against his body as though the planet is gazing at him and deliberately causing each of these mishaps. In many ways, Mars is. The planet is uninhabitable without myriad synchronized technologies to simulate Earth and constantly maintain the barriers separating these approximate environs from natural Mars. This is prior to any consideration of deus ex machina, in which Andy Weir is working within the planetary survival genre. The very nature of the red planet and the book will not permit Watney to have a fine time. The novel has two planetary settings, Earth and Mars—I will take up the intermediary location of the ship that carries the crew between planets in the next chapter. These two locations are structurally linked to literary points of view. Events on Earth are told by a third-person limited omniscient narrator while Martian happenings are in first person, relayed by Watney's log. Parsing the ability to gaze at the events on Mars by the characters as well as through the narrative structure and its omniscient deviations describes interactions with the universal ecosystem, entrenched and filtered by planetary environment and anthropos.

The Martian lends itself to interchangeably describing Mars at planetary and local scales—a familiar metonymic synonymity in the Anthropocene— because so much of Watney's survival concerns the Martian atmosphere. Exposure to Martian atmospheric pressure and temperature would kill Watney long before suffocating from the lack of breathable air. Watney is indoctrinated into this immediately upon waking and finding his suit and body punctured and reacting with the atmosphere. The event is fascinating because it is novel. Organisms must be in their environment, yet Watney cannot. Watney's biotic survival is most at odds with the total planet when he experiences the novel's two storms, as they ecometonymically trace from regional to planetary interactions. Weather, climate, describes and is manifested at the confluence of all planetary spheres. Focusing on storms in *The Martian* aligns with our top-down gathering of knowledge about the red planet, studying the outermost conditions before experiencing the surface. Gazing at Mars reverses historic human relation to our biotic home. We understood local biotic circumstance among organisms and landscape before codifying the relationship of global weather patterns and continental contours.[11]

The first of the novel's two storms scrubs the mission and separates Watney from the crew. NASA informs them of this approaching storm that could topple the MAV, the vehicle that gets them from Mars to the ship on which they will return to Earth. NASA aborts the mission and the crew begins to board the MAV. They walk from the Hab to the vehicle, blinded by the dust storm. Watney is struck by the antenna and separated from the crew. Unconscious, Watney remains on Mars buried over in dust. Without such forewarning, Watney faces the second storm alone as he journeys from the Hab to the vehicle that will get him off-planet. The dust preceding the storm is hardly noticeable, and if Watney travels too far into the storm before realizing it is on top of him, his solar panels will not be efficient enough to get him out. Entrenched in the Martian ecosystem, Watney must figure out that he is in the storm and how to get out and around before it blocks out the sun, halting the solar-powered rover in the storm and dooming him. NASA watches the Martian weather for the crew, disentangling them from immediate Martian environmental contours, while Watney is biotically entrenched in the ecosystem, bodily experiencing the storms.

Monitoring the weather, be it Mars's or Earth's, via satellites creates a distancing, non-biotic relation to the ecosystem. While rich in utility, weather reports are biotically oppositional and disintegrating. We monitor Earth's weather with off-planet technology, from within the Anthropocene epoch, where all but the most extreme weather conditions challenge only those othered by Western civilization. Wear a coat, stay inside, stock up on groceries, leave Mars. In the Anthropocene, humans use technology to keep the environment from impacting them. When terran communication fails though, Earth and Watney might as well not exist to each other. When Watney enters the storm, the satellites are unable to see him through the dust. NASA will either see a rover containing a dead Watney when the storm abates or they will see him begin to emerge from its edge to maneuver around the weather system. Antithetical to the rigid implementation of engineering, testing, and planning that makes NASA so successful and reliable, everyone is impotently reduced to hope: "Maybe the storm will dissipate unexpectedly. Maybe he'll find a way to keep his life support going on less energy than we thought was possible" (Weir 2014, 287).[12] For all but Watney, everything goes according to plan and the earthly microcosms maintain the separation of terrans from Mars. When disconnected from the earthly perspective, all that remains is the environment and the organism, the Martian storm and Watney, and their biotic relation.

A storm, Mars, once again consumes Watney and biotically entrenches him in the landscape. The first storm punctured Watney's suit and body, forcing an organic interaction of blood and atmosphere that seals the suit well enough for Watney to survive. Facing this second storm, Watney interacts with the Martian wilds as a more agential, albeit less visceral, cyborg. In a navigational misstep, he hits head-on the massive crater he intended to angle around. Putting a bit of effort into figuring out which direction is shortest, he exits the rover and walks up to Marth Crater's crest, noting the "stunning panorama" (Weir 2014, 295). The unintentional inaccuracy of his sextant-based celestial navigation forces Watney out of his rover to gaze upon an impact crater, a site evidencing motion and exchange in the universal ecosystem. He cannot see the other side of the crater as he expects. He turns around and the view behind him seems sharper, crisper, clearer, and Watney is unsettled by this "lack of symmetry" (Weir 2014, 295). The atmospheric environment in front of him is not acting as that behind him. Nature is not acting as nature. Well, nature is appearing asymmetrically. Storms are natural. Realizing his dilemma, Watney drives equidistant in both directions to deposit solar panels and return to the center. Establishing this array of solar cells allows him to measure sunlight in each location. The panel that gathers the most energy describes the quickest way out and around the likely circular storm. Technologically reacting to the sun, he moves toward its greatest concentration—a trick that even plants perform—to ensure that the sun continues to maintain his earthly microcosm. Formerly metonymic, Watney makes the technology that manifests the Earth in microcosm an extension of his senses, measuring more thoroughly what he visually acquired. Watney extends technology that produces power from gathering sunlight as a sensory array, continuing to integrate his biotic self and Martian landscape by technology for simulating earthly ecosystems. Extending this cyborg self—part technological Earth system, part terran body—into the Martian wilds is an amalgamating and prophetic site of human interaction with the universal ecosystem.

Watney accidentally comes upon this impact crater from some chunk wandering across the universal ecosystem to collide with this exact place. The mistake forces him out of the rover, as close as anatomically possible to Mars. Watney's awareness of baseline visibility, a notion derived from experience in the landscape, unsettles him. So, he compares his prior environs to his present and is not suddenly awash with realization. Instead, Watney is unsettled by a dissonance that he understands but cannot yet articulate. He senses rather than recognizes the problem. Luckily, "Watney is now an expert at surviving on Mars" (Weir 2014, 287). Survival is an act of proximity and awareness, of biotic interrelation with the environment. Watney's crewmates do not survive in the wilds of Mars. Watney does. He converts his instinctual reaction to uncertain stimuli into a logical description of the problem and solves it by extending his technosensory reach into the Martian landscape. And yet, for all his integration with the planet, Mars is trying to kill him.

In truth, Mars is not trying to kill Watney any more than Earth is trying to kill you. His phrasing, "Mars keeps trying to kill me" (Weir 2014, 229), conjures some Anthropocentrism, but the underlying process is identical throughout the universe: natural selection. Every environment challenges an organism to survive. That being's physiological equipment will provide some degree of capability relative to the environmental characteristics. Watney's blood happened to be suitable for patching up the punctured suit in the Martian atmosphere. Helpfully, the process worked even while he was unconscious. Mars only seems more deliberate and incisive because it is not our biotic home. Mars is trying to kill him as much as Earth tries to kill him, he just happens to be a species that evolved on this planet. Cyborgian adaptations have made us the likeliest terrans to adapt to Martian environs. Look how well—or poorly, the Anthropocene is pretty rough—technology worked for us on Earth. Organisms either survive or they do not. When they do, those survival tools are conveyed generationally in DNA or culture.

The entire novel seems to pit Watney, and all humans, against Mars as it relentlessly assaults the very biotic identity of the species. Andy Weir writes the earthly events of *The Martian* in third-person limited omniscient narration, creating a human collective that stylistically invokes the reader as another human watching Watney. However, the rare appearance of third-person omniscient narration inculcates us into natural selection in the universal ecosystem—which is the same natural selection on Earth and what is, from a certain point of view, trying to kill Watney. Driving down a natural ramp, Watney flips his rover by catching his wheel in a hole in the ground that has been filled with dust and blended by the wind to appear identical to the rest of the terrain. (The wind takes sides on Mars too.) Weir includes in the scene a narrative spanning millions of years as the Martian wind abrades the crater wall into a slope and deposits sand that makes the hole indistinguishable from the sturdy terrain. The rover is just the most recent event of the ramp. As that covered over hole implies millions of years on Mars, so too can the rover be metonymically traced millions of years backward. Follow the rover tracks back to the Hab which came from nowhere on Mars, find a local planet— local because these beings seem not to stay long but care enough about the lifeless rock to visit. The environment of the planet would be replicated to some degree by the rover and Hab, so that narrows it down to that one with all the water, right over there. Weir similarly traces the entire life of a piece of Hab canvas, from its construction all the way until the failure that explodes the Hab and makes potatoes extinct on Mars.

The omniscient narration allows the reader to track events through the universal ecosystem. Unsurprising, we have not yet considered ourselves as a part of it. Humans may not yet even be certain we have to be a part of the Earth system. Hopefully this book helps correct that. The Earth is but a mere patch of the universal ecosystem, where the simplest interactions are described by natural selection. Everywhere, all organisms are subject to natural selection, ecometonymies of their current and home environments. Tracking, as above, identifies these interactions across the universal ecosystem. Anywhere an organism leaves tracks or sign, on Earth or Mars, one can witness biotic identity. Tracing the rover's tracks is no different than tracing Watney's footprints. Such ecometonymy describes Watney's combined biotic circumstance, an organism entrenched in an ecosystem other than its biotic home—an imminent and novel circumstance for biotic life in our solar system.

APOLITICAL HUMANS IN THE
UNIVERSAL ECOSYSTEM

Many would argue there is no apolitical human action. If we call any action produced or reacted to by human civilization, sure, there is no apolitical human action. The anthroposphere connects to all planetary actions and is inherent in any invocation of the Earth system or its individual spheres. The Anthropocene, anthropos, frames all. This dangerously approaches, perhaps even conjures and institutes, the category of nonanimal human. Humans are animals, biotic organisms interrelated with their environment. My focus on the biotic human speaking as nature seeks to undermine and destabilize that all actions are politically informed and manifested by detaching human action from political initiations and extensions. Articulating processes of expansion to other planets codifies whatever becomes our planetary ecoethic at the end of the Anthropocene. Currently, expansion into the universal ecosystem is scaled up from our expansion into prior novel earthly ecosystems.

Philip Smith reads Watney's circumstance and actions as definitively colonial. I agree that any iteration of human presence of Mars can easily become a one-to-one transposition of "what has historically proven to be the manifestation of a violent colonial process" (Smith 2019, 332). Smith's reading is a valuable perspective for inoculating against the repetition of colonialization that perpetuates in the Anthropocene in myriad forms of ecological violence. Smith identifies that "*The Martian* and other contemporary writing on

Mars is a direct response to modern anxieties concerning humanity's future on Earth" (Smith 2019, 332). Thus, our interpretation of *The Martian* is exactly as determined as the character of our extraplanetary futures. Reading Watney's survival arc with the reciprocating ecocritical method to instead inculcate human biotic identity informs an integrative ecoethic actively resisting another colonial process. Colonial violence against nature and beings is historic and evident in our presence. I seek moments in which constructing the biotic human supersedes anthropos so that, when recirculated with human culture, the late-stage Anthropocene diverges from our consumptive trajectory. Biotic and apolitical, Watney informs the Anthropocene as the Parhams, the Child, and the kid—each demonstrating a progression toward an integrative ecoethic for the late-stage Anthropocene. Watney is entangled in familiar anthropos and embodies a biotic being after and within. Being the only human stranded on Mars disables earthly anthropos as opposite nature, even while invoking the tools of the Anthropocene.

Examining Watney as a terran biotically entrenched in the Martian environment describes an apolitical biotic existence excised from the concerns of anthropos and its dominant Western identity. Even the colonization implied by growing nonindigenous crops deteriorates when performed on Mars in this survival scenario. After regaining communication with NASA, Watney again receives data dumps of emails. Watney writes that he technically colonized Mars, as his alma mater reached out to say, "once you grow crops somewhere, you have officially 'colonized' it" (Weir 2014, 147). Of course, anthropos defining governmental and capitalist agendas get Watney to Mars and his accomplishment represents massive rhetorical and literal capital for his former university and NASA. Watney appears perfectly slotted between the two, but his actions are always driven by his biotic identity. He never agentially inhabits the middle space between these identities for his actions are biotic, as well as philosophically and temporally disconnected from Earth. Any political identity is manufactured on Earth, and never supersedes his biotic self. In fact, the politicization of Watney promotes returning him to his biotic home and has depoliticizing, at least depolarizing communal globality, effects on national space interests (Weir 2014, 195). Watney's distance from Earth exclusively creates him as a biotic agent entangled with earthly and Martian nature. While farming potatoes, Watney's connections to Earth are ecometonymic and through the maintenance in microcosm of an approximate Earth system. Focusing only on his biotic self, his anatomy demands he grow the plants, which he was not aware would constitute colonization—which here, in *The Martian*, becomes the least charged iteration of the term, not to mention his crops were never intended or even capable of sustaining a population for a time period constituting human permanence on Mars—nor would notions of colonization promote, inhibit, or otherwise alter the growing.

Watney's actions are biotic survival. His plans merely extend his life long enough to figure out how to extend it further. His survival is for the sake of survival, biotic and apolitical.

Human colonization on any planet conjures oppressive and destructive actions against indigenous environs and beings. Watney, the mission's botanist, studies the effect of Martian gravity and soil on terran plant growth. These experiments model the smallest terraforming iteration that SF regularly scales into turning the red planet blue for human habitation. Terraforming Mars is a vibrant discourse circulating imaginative scientific process and deep ethical perspectives (Gerstell et al. 2001; Fogg 2011; Schwartz 2013). Watney's experiments do not necessarily lead to humans converting Mars into a second Earth but do imply pursuit of Mars-based resources and self-sustaining colonies that consume Mars's chemical, spatial, and dirt resources, if nothing else. Stranded on Mars, Watney is not subject to such earthly political identities as his entire self is a biotic expression. His crops only extend his life, so he adopts farming practices that would strip the soil of nutrients in under twelve years, but he will only stay for a third of that (Weir 2014, 20). The soil would likely recover, except this is Mars. It will be Martian soil once the abandoned Hab decays. The soil will be desiccated by the Martian atmosphere, (dis)integrated and indistinguishable from Mars. Watney resides outside of the normal exchanges of anthropos and environment.

Watney's apolitical, biotic circumstance is enforced by survival needs but also manifests as his and the novel's base-setting. Riddled with postcolonial interplanetary circumstance, the potatoes traveled to Mars in natural, able to germinate form because NASA included a full Thanksgiving spread that the crew would prepare together. Watney is dismissive of the intended psychological care this provides. Of course, psychology and community are biotically human. Watney needs food with an urgency that deprioritizes and delays psychological distress. The former is mitigated by his work and routines while both are alleviated by writing his first-person narrative (as well as TV and music). Watney disassembles the Thanksgiving dinner brought from Earth, never again lingering on the components, except the potatoes he plants. Watney's characteristic joking dismissal guards against negative psychological impact, while displaying a disconnect from the celebration's inherent inclusion. He dismisses the ritual and uses a component part to convert the Hab into an organic Earth system that will sustain his biotic self. Watney is freed of a referent to colonization as oppositional to the land and indigenous beings, instead becoming kin with the bacteria, plants, and soil. Apolitical relation to Mars continues as he moves outward from the Hab and into the wider planetary landscape.

Watney's use of celestial navigation describes how locating one's position on a planet via the constant features of the universal ecosystem establishes and recognizes the inherent connection of the biotic self with immediate environs. Watney leaves the Hab on multiple long-range journeys in the rover. He utilizes maps and the landmarks noted thereon but was unable to use a compass because Mars has no magnetic field (Weir 2014, 97). Phobos, the largest of Mars's moons, travels east to west in such a tight orbit that Watney can view it twice a day to approximate cardinal direction and get within sight of the charted landmarks. For his longest journey, he needed a more reliable method to "fix his position on Mars" (Weir 2014, 288). To true himself, Watney constructs a simple sextant. While he charges the rover's batteries, he calculates longitude based on Phobos setting in the west. He has no problem overcoming the historic longitudinal problem of keeping accurate time and just plugs his data into a formula. Latitude is even easier. Each night, he sights Deneb with his sextant, the star aligned with the roughly 25-degree tilt of Mars, and calculates his latitude. He can never calculate lattitude and longitude simultaneously, instead using them to initiate travel direction the following day. Landmarks and terrain consistency additionally contextualize his location as he progresses across several geological regions.

Deploying even a simple sextant for celestial navigation, Watney defines his planetary location by the persistent features of the universal ecosystem. Latitude and longitude precisely describe one's position on a planet by understanding that world's relation to everything else in the visible universe. The concept of truing oneself gathers an otherworldly and magical quality by celestial navigation, as though our place on Earth is always our place in the universe. Tracing biotic identity in the universal ecosystem does not sever relation to the biotic home, instead detailing ecometonymic relation of self to local and universal nature. In David Barrie's history and personal use of the sextant, he explains that "the practice of celestial navigation extends our skills and deepens our relationship with the universe around us" (Barrie 2014, xx). The sextant provides "accuracy and reliability coupled with a deep immersion in the natural world" (Barrie 2014, 266). Barrie relates through personal and historical narrative that navigation goes beyond plotting coordinates. Voyaging requires deep attention to the immediate environmental circumstances as one moves in and to places. Contained in celestial navigation is the traceable interrelations of individual, local, planetary, and universal environments.

While once a Western development that secured shipping and colonization, the sextant now maintains only its more naturalist notes. Barrie argues that Global Positioning System (GPS) technology overtaking celestial navigation "banishes the need to pay attention to our surroundings and distances us from the natural world" (Barrie 2014, xx). GPS tells one's exact location via satellites humans placed in orbit, while the sextant connects individuals to the immediate environment and universal ecosystem. Requiring knowledge that takes generations to compile, navigating by sextant demands a planetary and universal awareness of the environment that looks more like natural navigation than technological positioning systems. Indigenous peoples of the Pacific Islands practiced natural navigation to travel consistently between islands. They knew where certain stars would rise on the horizon, which shift seasonally, and Barrie posits that the Pacific Islanders may have known which stars, at their zenith, would sit above important islands. Reminiscent of tracking's awareness of the landscape, natural navigation reads the

> distinctive patterns of waves and swells, which revealed to [the Polynesian voyagers] the direction of land long before it was visible . . . Of course they also carefully observed the behavior of birds, the nature of the clouds, and changes in the color of the water . . . sometimes even the taste of the sea could help them fix their position. It was extraordinary skills like these that enabled people not only to settle nearly all the islands of the Pacific, but to develop and maintain a cohesive culture embracing this vast area of ocean over many centuries. (Barrie 2014, 264)

Planetary ecology is entangled with the universal ecosystem at all scales and informs biotic identity.

Circulating even in pop culture, "We Know the Way," written and composed by Opetaia Foa'i for Disney's *Moana* (2016) describes the voyaging tradition as affirming the human in the environment as enabled by the stars.[13] The song describes natural navigation and identity:

We read the wind and the sky when the sun is high.
We sail the length of the seas on the ocean breeze. (Foa'i and Miranda 2016)

The island voyagers, in a magical sort of flashback witnessed by Moana, sing of their voyaging methods. Entwining the sun, the ocean, and the wind, the universal ecosystem is simultaneously issued with the hydro- and atmosphere and combined in the indigenous natural navigation for voyaging in the South Pacific. As with Barrie's description, albeit more magical, Moana manifests her cultural heritage across vast spatiotemporal distance. At night, the voyagers sing that they name each star, and by that knowledge,

We know where we are.
We know who we are. (Foa'i and Miranda 2016)

Natural navigation by the stars allows them the simultaneous claim of singular planetary and environmental location, and unified cultural identity. They connect the universal ecosystem to biotic relation with their environment. I am excited my daughter loves the movie, and I am excited to see a generation of children become leaders after seeing an indigenous culture's ability to biotically describe themselves by the universal ecosystem. Just as with tracking—related by Moana when she sings that "Every trail I track" brings her back to her biotic identity in nature as circulated by the act of voyaging from island to island by celestial and natural integration (Miranda 2016)—natural navigation further establishes indigenous sciences as a powerful ecoethical framework for countering the negative potential of technological development and lingering colonial processes in and beyond the Anthropocene.

In his forced, survival-based biotic relation to the Martian ecosystem, Watney models entrance to the universal ecosystem that is biotic, preceding and superseding political existence—an ecoethic that can emerge from a politicized present, utilize its tools, and yet maintain its priority of nondestructive relation to the nature that precedes our arrival. Weir is hopeful in this process. Watney is rescued with a Chinese rocket donated to the mission to save him. The discourse is kept among scientists because this cannot be resolved diplomatically (Weir 2014, 195). (It is politically framed and negotiated. This is still Earth.) Quickly though, Watney's earthly body biotically relating to Mars inspires a positive global human politic. Watney, back on Earth, reflects on his survival and rescue: "Part of it might be what I represent: progress, science, and the interplanetary future we've dreamed of for centuries. But really, they [saved me] because every human being has a basic instinct to help each other out" (Weir 2014, 369). Watney and Weir make a biological social argument that models the recirculation of the biotic human concept into the late-stage Anthropocene. Our grip on these apolitical moments is tenuous, if at all constant. Likely, the apolitical human is merely a spark that flashes into and out of existence with identical rapidity. Treated as nature narrative, Watney is just such an ember exemplifying a biotic human diverging from the destructive imbalance of the Anthropocene. Biotic identity must be invoked to institute an apolitical self which then reissues ontology and an ecoethic for the late-stage Anthropocene.

TRACKING WATNEY

With some satellite realignment and no storms, NASA can watch Watney on Mars. As they watch, they maintain their disconnected earthly perspective even while having a certain view of Watney on the planet. They never see him, but rather his metonyms: the rover, his suit, and, once they know he is alive, the Hab. All are Watney living firmly entrenched in Mars. The disconnect makes a strange border. Those on Earth can see Mars and the barriers against Mars maintaining Watney's microcosmic Earth systems. By the same barriers, no terrans see Watney's body or the simulated Earth environs. These are now curious extensions of himself, as cyborg or ecometonym. One says Watney but might actually be talking about the rover. Never privy to the earthly microcosm, those on Earth are disconnected from Mars and representations of Earth on Mars. The disconnect hilariously manifests when NASA watches each night as Watney deploys the modified emergency tent on his long journey. NASA agrees that this workshop space is a great idea, affording him greater room and mobility to maintain his critical technologies (Weir 2014, 276). Watney, however, wants to be able to stand up without putting on a suit and going outside, using it as a bedroom for comfort (Weir 2014, 281). The body needs little things, too. Only when tracking Watney in the landscape do terrans approach the Martian environs and experience Watney's biotic circumstance.

Tracks ecometonymically describe the precise organism in its environment anywhere that a track can be made. Deciphering Watney's survival, NASA deploys a noticeable lack of tracking in the dirt. The evidence they gather is all sign: two emergency pop-tents are deployed right next to each other and detached from their rovers, solar panels are cleared of dust, Watney's body is nowhere to be found, one rover is facing an otherwise impossible direction, and the landing struts left behind by the crew have been disassembled, which would never happen prior to launch (Weir 2014, 53–56). While discussing the signs of Watney's survival, NASA personnel make no mention of tracks, neither pointing to them as proof nor marking their absence among the several offered counterpoints. I suspect there are plenty of tracks in the scene, if not visible in the satellite imagery.

The timeline is a bit vague but it seems that Watney is maneuvering on foot and in the rovers within days of the images evidencing his survival. Returning from his first multiday trip in the rover, Watney backtracks the rover tread-marks to the Hab. These tracks are seven days old while those made sixteen days before are wiped out. Watney explains that even mild Martian weather would completely deteriorate the tracks in that time (Weir 2014, 105). There are probably still tracks in the dirt. Only four people view the images prior to announcing Watney's survival, none of whom are responsible for analyzing such minutiae, assuming the image quality would even register such markings. The signs are well read without needing tracks anyway. NASA characteristically focuses on the story told by the technology that enables the crew to live despite the abiotic features of the Martian landscape. They see the things they built to help simulate Earth against Mars, maintaining their literal and perspectival disconnect from organic Mars. Inability to read the tracks is a nonissue once NASA reestablishes communications. When they again lose comms, Watney updates them with messages spelled in rocks on the planet's surface.

While NASA does not yet look at tracks on the ground, Watney is thoroughly entrenched in biotic interrelation with the red planet, keenly imbibing the baseline Martian terrain. Above, Watney offhandedly narrates his understanding of Martian weather's baseline relation to track degradation and total erosion. Following his own tracks until he loses them brings him back to local terrain and familiar landmarks by which he employs natural instead of celestial navigation to arrive at the Hab. Watney's tracking knowledge is absorbed through consistent proximity and attention to the immediate landscape, deploying scientific processes without experimental rigor or even overt attention. Watney gathers the ecological baseline by living within, becoming an excellent Martian tracker.

Preparing for Martian surface expeditions, Watney scientifically assesses the rover's baseline performance on Mars. The first test measures the rover's baseline travel efficiency over Martian terrain and how long he can last with the heat off (to convert life support power into distance). Both tests are failures. Watney repetitively drives the rover back and forth over the same stretch, quickly wearing down a path that noticeably increases the vehicle's efficiency by modeling a compacted path instead of wild terrain. He begins to randomly alter course and simulate actual travel. Additionally, Watney hypothesizes that the evolutionary technology of being warm-blooded paired with the regular technology of multiple sweatshirts will maintain the rover's earthly climate against heat lost to Mars through the vehicle's insulation. Watney disables the heater, turning it back on after just an hour and returns to the Hab (Weir 2014, 72). Watney applies scientific rigor without incorporating the immediate environment. Nonbiotic, his tests fail almost immediately. Watney works through the problems to achieve biotic interrelation characteristic of his casual success in tracking and natural navigation. From there, further biotic experience with Mars culminates in the full integration detailed by surviving the two storms, manifesting cyborg-in-the-wild Watney.

After the final storm, NASA reads tracks in the Martian dirt to access Watney by his biotic ties. Looking at the images of the flipped rover on the crater ramp, Mindy, first to discover Watney is alive and now tasked with positioning satellites to constantly track him, explains, "'The weather's clear. If he'd come out, there'd be visible footsteps . . . there are ordinary wheel tracks' . . . she pointed to a large disturbance in the soil . . . 'Judging by where that ditch is, I'd say the rover rolled and slid from there. You can see the trench it left behind. The trailer flipped forward onto its roof'" (Weir 2014, 311). Mitch, head of the Jet Propulsion Lab, adds that the rover is intact for "if there'd been a pressure loss, there'd be a starburst pattern in the sand" (Weir 2014, 312). This is as close as NASA gets to seeing Watney without him exiting the vehicle, inside of which he or his life support may be irreparably damaged. Mindy and Mitch describe what *The Martian*'s third-person omniscient narration of the sand-filled hole grants the reader in the prior chapter. Tracking Watney grants the novel's Earthbound characters access to the same narrative omniscience of the reader.

AN EARTHLY CIRCLE OF NATURE

My tracking framework is grateful for the myriad track images in space travel and SF. The motif is admittedly unsurprising. Tracks describe travel, presence, and identity. One can reflect upon her own track as it describes an unprecedented personal and species relation to a novel environment, a first or last contact able to be looked upon from within the experience. Before boarding the heavily modified Mars Ascent Vehicle that reunites him with his crewmates, Watney reflects, "I've left my last footprints in the dusty red sand" (Weir 2014, 341). Regardless of the launch's success or failure, Watney's trackline on the red planet ends. Tracks are not the only sites of Watney's biotic relation to dirt. Leaving Mars required him to convert the Hab from a technological human container to a complex ecosystem simulation. Watney needs to make land to farm. Reminiscent of the ecological control characteristic of the late-stage Anthropocene, Watney creates near-perfect growing conditions for an interplanetary potato crop.

Watney calculates that this potato crop combined with the mission supplies can get him within range of the next terran crew to arrive on Mars. The crop will at least buy him time to figure out how to further extend his survival. Watney suspects that his crop will produce a higher yield than farms on Earth because it is feasible for him to give each plant individual attention and there are no blights, parasites, animals, foul weather, or weeds to hinder the plants (Weir 2014, 71). The Hab climate is set to optimum growing temperature and light conditions. Joking that he knows the recipe for water, Watney employs an incredibly dangerous chemical process with a high likelihood of exploding to create the large amounts of water needed to sustain the crop. Watney only brought enough Earth soil to run his botany experiments, small-scale assessments of the viability of Martian soil, and now must make more. He gathers available biomatter into a compost bucket, scraping in the limited leavings of his rationed meals. Watney adds his own waste and gathers the mission's desiccated feces that was bagged and deposited onto the Mars surface, killing any problematic microorganisms (Weir 2014, 13). Mixing this with Martian dirt creates enough arable soil to fill the floor of the Hab, unused bunks, and tabletops.

As the mission's botanist, Watney's experiments explore the much-theorized notion of growing plants in Martian soil. He states, "Martian soil has the basic building blocks needed for plant growth, but there's a lot of stuff going on in Earth soil that Mars soil doesn't have, even when it's placed in an Earth atmosphere and given plenty of water. Bacterial activity, certain nutrients provided by animal life, etc. None of that is happening on Mars" (Weir 2014, 12). To provide the necessary bacteria and fertilizing biomatter to the lifeless Martian soil, Watney rehydrates the desiccated waste and all the while adds in his own manure. He mixes the compost into the Martian soil, finishing by sprinkling the Earth soil on top because "There are dozens of species of bacteria living in Earth soil, and they're critical to plant growth" (Weir 2014, 14). He lets the bacteria multiply and make earthly the soil. Adding more Martian dirt in stages doubles the amount of soil until the Hab is covered.

Watney's extreme circumstance imbues this terraforming with a strange and antithetical disruption of civilization. Terraforming Martian soil with his waste, Watney remarks that the bacteria introduced by using night soil do not cause disease: "The only pathogens in this waste are the ones I already have" (Weir 2014, 14). Watney's farming method could prove deadly with additional humans. Night soil is "not an ideal way to grow crops, because it spreads disease" (Weir 2014, 14). I expect astronauts, who return from the space station with greater diversity to their microbiome (Gilder 2020),[14] would show the same overlap as any cohabiting adults. Close and regular proximity to other beings, even across species, creates marked similarities in microbiome character, but there remains enough variation in the microbiome for pathogenic possibility, even among cohabiting identical twins (Song 2013). Adding another human to Watney's proprietary ecosystem could produce infection that prevents already necessarily malnourished bodies from performing work to survive the otherworldly environs or cause death outright by dehydration and other deficiencies.[15] The terraforming action that, to my knowledge, exclusively implies colonizing civilization in SF fundamentally opposes human proximity in *The Martian*.[16] While it would likely offer additional survival options, the bacteria forming the soil become dangerous, pandemic terrors.

Instead, bacteria are critical kin, digesting and nutrifying plants to survive with and as interplanetary organisms. Divorced from civilization, the bacteria that inherently construct this Earth system traverse all things and places. While they interact with their surroundings by decay, everywhere in the simulated Earth system they are beneficial and promoted. In the Hab, the bacteria are never doubled in their environs, always acting as they are and securing the natural processes. They decay soil components so that plants may grow, just as they break down potatoes in Watney's gut for his survival. They exit Watney and enter the soil. They are eaten by Watney but never consumed, rather reintegrated into another biotic location and process. Watney simulates the Earth system's interrelated processes by recognizing the critical kin and kinmaking of bacteria from his microbiome. As Haraway says, "To be animal is to become-with bacteria (and, no doubt, viruses and many other sorts of critters)" (Haraway 2016, 65). As animal, Watney's biotic existence secures and is secured through simple biotic relation. Watney's biotic relation to the earth is simplified and intensified, which in the Hab is also an ecometonymic description of Earth and Mars in the universal ecosystem.

His in-extremis survival allows us to move beyond the problematic notions of his actions, as planned food production would just bring supplies that dissolve the need to employ less-sustainable methods. This grants insight into the ability to turn entrance into the universal ecosystem into a process of re-civilizing by embracing pre-civilization models and diverting from those practices that create the Anthropocene. Watney represents a planetary kinship at the site of biotic and abiotic function that coalesces the universal ecosystem and treats Haraway's interconnection model as an ethos instead of epoch. Examining Watney's relation to bacteria allows us to stay with the trouble even though we will leave the planet. Haraway tells us that "Of course, from the start the greatest planetary terraformers (and reformers) of all have been bacteria and their kin, also in inter/intra-action of myriad kinds (including with people and their practices, technological and otherwise)" (Haraway 2016, 99). Watney describes a production that can inform a proper relation to ours and other planets, to the universal ecosystem. After all, Watney contently reflects, "I don't mind walking on dirt" (Weir 2014, 14).

AN ASTROPIC ACT

Watney is a case study in the late-stage Anthropocene. His tracks relay integration with a planet the human body did not evolve on, the simulation and curation of an earthly environment off-planet, and how natural selection occurs at the ecological scale of the universe. Watney is a human-animal and his kin living on a planet other than their biotic home whose farming methods launch habitability efforts forward. The late-stage Anthropocene is characterized by control exerted on planetary ecology. Continuing to spread beyond our biotic home demands approximating Earth in microcosm, a knowledge that reciprocally informs and is informed by the ecological control exerted on the Earth system. In the late-stage Anthropocene, humans are constantly asserting and exerting ecological systems that simulate earthly environs.

Watney's extreme circumstance regularly fractures the barriers guarding these simulated earthly environs. He reacts in a variety of ways that largely converts abiotic Martian features into various maintaining barriers or versions of the Earth system. His original experiments imply and precede cultivation on Mars. Watney maintains that trajectory and launches ahead a few steps to grow a sustaining food crop. He multiplies the soil by combining it with Martian dirt, increasing the amount of product without increasing the original resources, which are hardly ideal for the task anyway. The multiplication of dirt is a consumptive act that makes it more akin to Earth soil than Martian dirt, a sort of proto-terraforming. The planets are still bordered, with the earth consuming and converting the Martian dirt. When the Hab breaches and is later abandoned, reciprocal consumption occurs and the soil is eventually made indistinguishably Martian, excepting some bacteria sealed in microscopic ice and buried over, although this is only discovered when Watney reestablishes the Hab's earthly environs. In this and all other actions, Watney biotically relates to Mars.

How he stores his harvest though shifts the character of his biotic relation to Mars to something asynchronous with the late-stage Anthropocene. Left in the Hab or rover, the potatoes would rot in the warm, pressurized environment long before he could consume them. Instead, he does "something that wouldn't work all that well on Earth: throw them outside. Most of the water will be sucked out by the near-vacuum; what's left will freeze solid. Any bacteria planning to rot my taters will die screaming" (Weir 2014, 150). Incorporating the natural Martian environment to halt decay, Watney enters a fascinating ecosystem in which the planet simultaneously denies and preserves his "simulate[d] complex global ecosystem" (Weir 2014, 13). With this action, Watney reciprocally invokes the Earth and Martian ecosystems. Mars envelops the potatoes, subsuming the natural decay inherently maintaining Earth systems. This in turn bolsters Watney and his bacterial kin's survival in the simulated Earth system. The Earth system now perpetuates by the oppositional function of the larger Martian ecology that has until now consumed or resisted Watney's "circle of nature" (Weir 2014, 79). This act is akin to Watney's initiating circumstance, where the Martian atmosphere evaporated the water in his blood, sealing his suit. Incorporating the Martian landscape to halt entropic decay and promote his body, Watney includes the unaltered Martian environment to halt and promote an earthly ecosystem, biotically integrating with another planet.

And with this or some self-same act, the Anthropocene ends.

NOTES

1. The imperial utility of the dark side is dominant in *Star Wars* but not inherent, and the extended universe imagines further iterations of dark side Force use. The Night Sisters in *Star Wars: The Clone Wars* (2008–2020) and *Star Wars: Dark Disciple* (Golden 2015) develop the complexity of the dark side initiated by the cave on Dagobah.

2. This may ultimately need amending as well. There are tardigrades and perhaps other microorganisms in discarded waste on the moon—albeit in various suspended states (Scharf 2019; Oberhaus 2019).

3. This is not meant to undermine ecocritics working with hypothetical but plausible exoplanetary or fantastical ecosystems. Working with fictional planetary environs is critical for constructing the ethos that guides our interactions with both earthly and extraterrestrial ecosystems.

4. I imbibed this imaginative architecture from the scene in Christopher Nolan's *Inception* (2010), where Cobb uses these arrows to explain how humans, in a dream, simultaneously create and perceive the dream world. I use it to describe all sorts of reciprocating actions difficult to separate but easy to initiate (you can start with either and it just continues on its perpetual work). The image describes all sorts of ideas from this ecocritical model to how narrative is comprised of characters and plots reacting to each other, just as organisms and environment.

5. Petranek's description of Martian exploration and colonization consistently and problematically invokes a motif of westward expansion and colonization. I worry about instituting this colonial attitude toward Mars while also noting how likely humanity is to act anthropocentrically Westernized even without Petranek's frame. Philip Smith (2019) further examines the parallels between Andy Weir's *The Martian* and the American frontier narrative.

6. I was of two minds when I first watched the series in 2018. I was annoyed to find that I was not first or second to invoke this reciprocating action of SF surrounding now viable missions to Mars, and enthusiastically validated. I knew it was real. And now, the method progresses ecocriticism and human paradigm.

7. Chapter 5 will recirculate the method to a broader literary framework, which it must engage if it is to be a paradigmatic technology.

8. This also looks very much like Timothy Morton's ecological thought as it witnesses the "mesh," albeit with my tighter focus on access points.

9. Limiting this analysis to these two does not dissolve any other SF visions, but rather draws out the texts this analysis prioritizes: those that reward the reciprocating ecocritical action. They need be no less speculative feminism or string figures.

10. Focusing on prosthetics and implants (like pacemakers or cochlears) seems to more directly frame concepts of cyborg elicited by suits and ships for the interplanetary biotic human than cell phones and the Internet. Of course, communication technologies will be close at hand and inextricable from extraplanetary existence. For Watney though, the technology is mostly attending to physiological and mobility-centric needs on Mars. That said, TV and music are there, too, and quite valuable.

11. Increasing our focus on indigenous sciences will certainly reveal a greater knowledge base arriving earlier than Western thought expects. Likely, the knowledge comes with a process for—because it comes from—biotic entrenchment, as with tracking and navigation. Such versions of thinking promote wonderful ecoethics for shaping the late-stage Anthropocene, and are objectively—fine, "in my opinion"—far cooler than the versions that are oppositional to the environment.

12. I would include a note from Gregory Sargent here, as he took the deepest dive into the manuscript and his effort and thought are as inextricable from this book as my own: "Yeah, this is an interesting point, especially since it's conveyed by using the subjunctive mood, which has something to do with your book's project of changing the narrative about the coming end to the Anthropocene. It's about switching tenses

in the way that humans do things." I love a structuralist reading and believe this comment to be a telling analysis of *The End of the Anthropocene.*

13. Amid understandable appropriation concerns surrounding *Moana* (2016), "We Know the Way" blends Polynesian language and English, without translating, and shows a vibrant culture deploying natural navigation to spread to other islands. The larger theme of Moana is that such expansion promotes the conversion of volcanic formation of the environment to harmonic environmental sustainability. All is accomplished by a young woman who manifests as an ecofeminist model of the feminine who positively and biotically maintains nature.

14. Though this is suspected to be caused by the general increase in dietary diversity.

15. The antibiotics in the Hab could make it a nonissue, but adding another person doubles the resource expenditure which halves the survival time frame. The longevity the potatoes provide is what allows NASA enough of a window to effect rescue.

16. This could be avoided by simply planning to create a garden and bringing enough fertilizing components, although weight restrictions inspire some very cool thoughts beyond just bringing dirt with you.

Chapter 4

The Astropocene

The Astropocene begins.

When there is no mandatory biotic tie to our home planet—when other planets do not refuse us—we begin the next terran epoch. By the end of the Anthropocene, humans acquire unprecedented control of Earth's ecology. In the Astropocene, we extend our reach beyond the planet, capturing asteroids and comets and gathering their materials. There are novel interactions with the universal ecosystem, and we leave a bunch of footprints in places no terran has ever walked. Terran identity is informed by biotic relations in unprecedented ways and environs, and it continues to develop the cyborg lines we already exist along. The Astropocene is about novel biotic interactions with the universal ecosystem.

The Astropocene inherits its character and trajectory from the late-stage Anthropocene. Humans in the late-stage Anthropocene leave Earth and enter the universal ecosystem to stay. Permanence will not immediately manifest as our SF mythos describes. Terrans living their whole lives in permanent extraterrestrial colonies orbiting the moon, on Mars, and in the asteroid belt is the work of generations.[1] We will push further from Earth for reasons ranging from the idea is awesome to planetary cataclysm. The impetus only determines the rapidity and preparation. Human trajectory in the universal ecosystem progresses from the International Space Station to the moon and to Mars (NASA 2020c).

Concurrently, humans shift their relation to planetary ecology from environmental impact to control. Controlling our environmental impact necessitates generating consistent productivity across scales. We could drastically restrict how much we impact the innate function of the planet, even aggressively reinstitute historic ecologies. Alternatively, we might maximize control and author the planetary ecology sustaining only those ecological parts we cannot efficiently replace, constantly tweaking planetary ecology. Of course, we arrive at some mediation that is already axiomatic in our environmental ethic. Reintroduction of species is dramatically embraced at several levels. Apex predators and herd animals are being reintroduced and many people prioritize native plants while landscaping their yards. Reforestation is valued and promoted, though it does not instantaneously repair biodiversity. What even guarantees displaced and extinct species return or have not already impacted other ecosystems in relocating? Terrans are likely stuck with a best-case scenario where the planet only gets a bit warmer before we use the anthroposphere to stabilize the Earth system. Such efforts do not revert the Anthropocene to the Holocene, nor do they end the epoch.

Living in the late-stage Anthropocene is wrapped up in the problems of being doubled on all planetary objects. We are wardens nowhere. Maintenance of our species must drastically change if it is to maintain the rest of the ecosystem. The problems of living on another planet become strangely familiar to the problems of living on Earth. In an odd reversal, one concept for terraforming Mars releases greenhouse gases into the atmosphere with the intention of melting polar ice (Gerstell 2001)—eerily securing a habitable Planet B by the means we brought Planet A to its brink. Such terraforming converts our planetary impact to a technology for controlling planetary ecology. At the end of the Anthropocene we are entangled in the universal ecosystem by spreading to live and die beyond our biotic home. Mark Watney is a version of such expansion.

Eventually, the universal ecosystem will not refuse us and the Anthropocene, the last planetary geological age, ends. The Astropocene is a paradigmatic confluence of geology and culture. The story of terrans will no longer be limited to Earth. There are already individuals among us whose bodies ecometonymically describe biotic interrelation with the extraplanetary universe. Astronauts return with persistent biological shifts "observed even after 6 months on Earth, including some genes' expression levels, increased DNA damage from chromosomal inversions, increased numbers of short telomeres, and attenuated cognitive function" (Garret-Bakelman 2019, 1). We are still uncertain of the permanence and impact of these shifts. Eventually, humans will have immediate biotic interrelation with the universal ecosystem. Which, of course, we already do. The sun and the moon contour the expression of biotic and abiotic features of the landscape. Geological epochs wax and wane according to the complex interactions of planetary systems: bacteria oxygenate the planet, or an asteroid impacts and the resulting dust blocks out the sun, or the Earth warms, forests recede, and fleshy running animals start doing their thing. A lot of dense stuff moves incredibly fast in the universal ecosystem. Shortly, we will exert control and go get those things for our own designs (NASA 2019a, 2019b; *U.S. Commercial Space Launch Competitiveness Act*).

We will no longer tell our biotic story, in any discipline, by this planet alone. When we are surviving by biotic relation to non-earthly environs, so begins the Astropocene.

In the next epoch, humans find themselves biotically related to the universal ecosystem, drawing on the resources it provides and shaping ecosystems as needed. The abiotic components of an environment can exist without the biotic, and the closest planetary objects are not going to tolerate terran organisms without technological mediation. Watney's use of the Martian atmosphere for food storage signals the necessary thought and action for existence in such novel wilds. Watney's astropic act signals interrelation other than the hallmark resistance to biotic extraplanetary environs. Our biotic home is already the universal ecosystem, we just tend not to think about it. Terran bodies form and function by abiotic factors like sunlight and gravity. These and other resources vary in the universal ecosystem and extraplanetary existence will depend on novel mediation of Earth simulations and surrounding environs to maintain the body.

IN THE ASTROPOCENE

"The moon blew up without warning and for no apparent reason," begins Neal Stephenson's *Seveneves* (2015, 4). A puff of dust on the moon's surface, likely a meteor impact, is first noticed by an amateur astronomer who makes a grab for fame by posting the event online. He was forgotten. The moon exploding is a relatively simultaneous human experience. If you are looking at the moon, you see it explode. If you miss it, there is a dust cloud where the moon was. In *Seveneves*, human chronology resets to this moment, called A+0.0.0 or Zero (Stephenson 2015, 3). Never demystifying the event, the novel explains that the leading theory is that some object pierced the moon's surface, burrowing to the center where it released its energy. Perhaps it kept going and simply deposited enough energy in its passing to break the moon apart (Stephenson 2015, 4). The characters are generally content with the idea that things can get moving incredibly fast in the vacuum of space and impacting something converts that speed into a lot of energy. More than once, *Seveneves* asserts the constant danger of even a tiny spacefaring object punching through the ISS or any space habitat hull. Depending on size and speed, it could even be undetectable. Additionally, there's not much that can be done to unravel the event, especially as avoiding extinction by the "Hard Rain" of moon debris eventually falling to Earth commands attention.

Capturing human relation to the universal ecosystem, *Seveneves* details that "an unknown Agent acted upon the moon. The moon, along with all the humans living in the sublunary realm, was the passive recipient of that action.

Much later, humans might rouse themselves to take action and be agents once again. But now and for long into the future they would be nothing more than patients" (Stephenson 2015, 5). Applying ecocriticism to read *Seveneves* as science fact fiction prioritizes the most scientific hypothesis: the moon was hit by some natural object traveling at tremendous speed. Affirming the eco-critical reading, the effect is biotic and deprioritizes anthropos and ontology. The moon breaks into seven large pieces, maintaining proximity and thus roughly the same gravitational center, only marginally impacting the moon's gravitational relation to the Earth system. Eventually, collisions further break apart the moon's pieces, giving Earth some very cool rings. The gravitational impact is hardly important during this White Sky period as the decentralized masses are pulled to Earth. This cataclysmic Hard Rain lasts thousands of years.

Stephenson defines agent in a martial context, explaining that in a sword-fighting drill the agent is the attacker while the patient is passive or respond-ing. This creates in the event a passivity to the moon and to humanity, a natural violence reminiscent of Mars trying to kill Watney—again, it is but as natural selection. The effect has scaled to the interactions of the universal ecosystem as they affect the species instead of Watney as a representation of humanity. The (decon)structuralist in me must mention, unlike the novel, the agent-patient relation is roughly the same in grammar, with the patient being acted upon by the agent. The novel and its implied editor(s) eschew this connection. Doing so, the image strictly removes the more civilized metaphor favoring a visceral and penetrative moment that, by biotic pathways, affirms humans are always already included in the interactions of the universal ecosystem—after all, "There's literally everything in space" ("Rattlestar Ricklactica" 2019). Intricate grammar as an extension of civilization is unsuitable as a metaphor for the reengagement of the ecoreal, and the desired reader response. The inclusion of the sword, of steel technology, deepens the Watney-like cyborg identity humanity must manifest to survive in space for thousands of years and foreshadows latent tensions of colonial and non-biotic violence.

Signaling for terrans their already existing interrelation with the universal ecosystem, the moon explodes "And the universe changed" (Stephenson 2015, 12). This change is not that humans have left their biotic home. We would have done that anyway, with any immediate actions unlikely to regis-ter at the universal scale. According to *Seveneves*, the abiotic processes of the universe are expressed by planetary interaction. My insistence on scaling the ecosystem to the full universe instead of the solar system or galaxy is derived from such concepts. The Agent traveled from precisely anywhere, probably jettisoned in some similar collision, perhaps even continuing on out the other side of the moon to wander further. Pieces of the moon are likely now on such courses too, traveling for what millennia to impact something or nothing. The impact of moon shrapnel crashing to Earth registers in gravitational character and planetary orbits.

The chaotic happenstance of the universal ecosystem strips humans of their biotic home to become cyborgs in the wilds of space, with only technological habitats biotically extending to describe humans. After the Hard Rain, humans are not agents for a long time. In *Seveneves*, humans gather extreme control over ecology and engage the universal ecosystem to that end, but do not grow beyond the ability to shape planetary ecology by the end of the novel. The characteristic control of the late-stage Anthropocene is subverted by the inability to impact the gargantuan scales of the universal ecosystem. There is no way to exert control over this environmental shift. The bulk of the novel then follows the survivors who make it into space eventually settling on one of the large lunar chunks. The Anthropocene ends with total terran species living on that rock.

Suddenly, 5,000 years later, we join Kath Two on the terraformed second iteration of Earth. Her scientific mind at odds with the innate human desire to see the beauty of nature, Kath Two tells herself the "water in that lake below . . . is there because we crashed comet cores into the dead earth until it stayed wet. The air was manufactured by organisms we genetically engineered . . . then killed once they accomplished their task. And the sharp scent . . . comes from vegetation that, for many years, existed only as a string of binary digits stored on a thumb drive. . . . None of which changed the fact that she liked it" (Stephenson 2015, 575). In a thought, Kath Two describes the reconstitution of Earth. As Watney invokes the Martian atmosphere to simultaneously halt and promote the Earth system, these people reverse their cataclysm, crashing comets to Earth to revivify bacterial life. Plants and animals follow once the atmosphere takes hold. Using comets as a natural resource to preserve and simulate earthly ecology is the very stuff of the Astropocene. In crashing comet cores to rehydrate the planet for life, the species divorces its actions from the anthropocentric concept of cataclysm, instead controlling the circumstances that direct the natural resources of the universal ecosystem. If comet cores are unneeded for Earth, I would not be the first person to mention their utility in terraforming Mars (Kaku 2010) or in manufacturing rocket fuel and natural resources for traveling between these patches of the universal ecosystem (Yeomans 2016, 13).

The Anthropocene ends as humans leverage their ability to control aspects of planetary ecology for biotic interrelation with the universal ecosystem. We were always subject to it, as *Seveneves* apocalyptically details, but we generally only consider the universal ecosystem when something from it comes to us or when we go elsewhere. The only difference between the transition in *Seveneves* and what I imagine is the extent of the cataclysm. Even in the Anthropocene, with the sixth great extinction potentially at hand, such destruction is not needed to initiate the Astropocene. Technology in *Seveneves* is only slightly ahead of our current abilities to harvest passing objects and bioengineer organisms for insertion into an environment (Trivedi 2011). The same mechanism of control by which we relate to our biotic home we use to relate to the universal ecosystem. The impact on our planet and all other planets reciprocates as we begin to terraform our world or other worlds and spread into the universal ecosystem.

TERRAFORMING AS NATURE

In *Seveneves*, humans are excised from their biotic home but never manage to divest themselves of innate need for the natural environment. The Earth system is described by their bodily self, which is prioritized in vacuous environs and inorganic spacecraft. They are ecometonymies. Their biotic identity is reciprocally constructed, and thus always implies the contours of the environment in which an organism lives and evolves. Excised from the planet, that genomic structure instead becomes an ecometonymic hyperobject that describes both the organism and its nonexistent environs. For *Seveneves*, the human genome continues to ecometonymically locate human biotic identity in terms of Earth's planetary ecology. Just as no form exists without the dimension of duration, nor can an organism be without environment. In *Seveneves*, the genetic response formed by earthly environs is never selected out. Even after extensive genetic engineering and 5,000 years of living in a space station the extant features evolved by earthly interrelation remain: "if it doesn't kill you you can keep it" (Morton 2010, 85).

Responding to encoded primordial relation that manifests as instinctual desire, humans "had built habitats large enough to support lakes and forests. But nature simulated was not nature" (Stephenson 2015, 572). Let us assume that these habs perfectly reiterate earthly environs. The genetic makeup of the organisms is exactly replicated from Earth DNA; the environmental sciences has gathered enough observations that they can describe and implement perfect and balanced biodiversity; we can even say that the technology to produce exactly 1G of gravity exists, implied by the other abilities. What is missing from such total ecological control? How is it a mere simulacrum? This nature lacks connection to everything else, is disconnected from a planet and the universe that grant fundamental components and interactions. There are no transitional spaces where the animalia and plantae interact with one another. The environment ends. No wind and weather circulate across continents. No migrations occur. The place is static, anachronistically seasonless and devoid of any notion of planetary ecology and thus is not an ecosystem. The forest simulacrum implies the planet metaphorically. The ecometonymic pathways halt at the walls and doors of the spacecraft, referring to the human record of its pre-cataclysmic planetary system. Earth is known by humans and then expressed by humans. Even if perfect, it is still not an ecometonymy. A forest implies the environs at its borders and the animals who pass through depositing plants and microorganisms along their migratory route. This is the very knowledge that shifts human impact to control, initiating the late-stage Anthropocene. *Seveneves* distinguishes between this habitat and the game preserve that is the terraformed New Earth. Their plan was always to return to their biotic home planet.

Humans may be able to create nature, but they cannot be extricated from nature. Terraforming New Earth, humans bombard

> the parched and dead surface of the planet with comet cores for hundreds of years just to bring the sea level up to where they wanted it. Then they had infected the water with organisms genetically engineered to produce the balance of gases needed to support life—and having done that, to commit suicide, so that their biomass could be used as nutrients for the next wave of atmosphere building creatures. According to their measurements, the result was a nearly perfect reproduction of Old Earth's atmosphere. (Stephenson 2015, 656)

In a series of astropic acts, invoking the resources of the universal ecosystem for the maintenance of Earth systems, humans replicate the prior planetary system. They begin by gathering comets to the planet. Humans in the Astropocene are very aware that we are organisms in the universal ecosystem. All precedented life demands liquid water and the icy cores are a natural resource abundant enough to saturate the planet and reconstitute the oceans. Then, exerting genomic control, they design microorganisms to convert the resulting molecular identity of the planet to the balanced composition necessary for supporting earthly life. These organisms die on schedule and nutrify the planet for the forthcoming, likely photosynthesizing, organisms that maintain the atmospheric composition. Implied in this balance is a pre-Anthropocene planetary ecology. While not explicitly stated, the very ability to create a self-sustaining atmosphere with comets and microorganisms implies a knowledge of historic atmospheric conditions and incentive to not reinstitute a polluted system of human-induced extinction. Doubled as they were by cataclysm, the humans of *Seveneves* were spared the Anthropocene.

The Astropocene is characterized by such unprecedented control over human relations to the environment. Conceptually, astropic acts are of an environmental scale. Here, humans have already expanded into the universal ecosystem, going from a clustered mesh of spacecraft acting like a school of fish to a static existence on a lunar fragment, to a civilization on a massive station orbiting the planet. Their new races are a product of genetic modifications selected by the titular Seven Eves to highlight what they thought must be perpetuated in their offspring, the species. They mined asteroids and deployed comets to terraform an entire planet with their mastery of bioengineering, making organic technology of bacteria. Such control seems to run counter my earlier assertions that humans are organisms incapable of extrication from nature.

Instead of being doubled in some Anthropocene paralleling Frankenplanet or alien doubling action of Area X, humans biotically confirm New Earth. Axiomatic to the novel, "No one who breathed [the replicated atmosphere] after a lifetime spent in habitats needed scientific data . . . Its smell penetrated to some ancient part of the brain, triggering instincts that must go all the way back to hominid ancestors living on the shores of Africa millions of years ago" (Stephenson 2015, 656). No astropic action disrupts the biotic notion of home described by an organism. Of course, space travel presupposes knowledge of breathable atmospheric levels but the body itself describes these things as well. Humans do not experience it quantitatively, but the biotic identity, genetically and physiologically, contains ecometonymic descriptions of the very environment driving our speciation.

Seveneves links physiological affirmation with the pastoral visual, a palindromic reversal of the emergence of the aesthetic. Williams tells us of landscaping as an art, "The genius of the place was the making of a place" (Williams 1973, 124), a notion with which I would adorn terraforming. Williams describes how the English countryside becomes a canvas for the landscaping art, making the land a constructed view. This pastoral form of ecological control philosophically precedes terraforming. *Seveneves* doubles the action by landscaping at the planetary scale. What is represented by the English landscapers is constructed as naturally as possible by terraforming. Each purport to recreate an authentic nature bodily, the sublime for English landowners and phenotypically for the terraformers. Human technology funnels toward terraforming. This progression grants the Astropocene options and vision to exert ecological control with a connection to the primordial environs that made us.

Invoking Chris Pak's feedback loop reiterates how the pastoral image became of value to industrial civilization, while terraforming to create a validating connection with our primordial selves is an expression of the human-environment dissonance of the Anthropocene. Truly, "When alternative views of the past are dialogized, the past and future begin to reciprocally shape each other" (Pak 2016, 212). Tweaking the temporality further enables the reciprocating ecocritical method.[2] By issuing science fact fiction as the past we witness its trajectory and affirm the characteristics of the Astropocene as they extend from the Anthropocene. This is the very narrative structure of *Seveneves*. The scientific expertise and genetic predisposition of the survivors directly manifests as the contours of their speciation and civilization. What is consistent is the return to the inherent value of nature as itself. Perceived separation and inability to control the environment institutes the Anthropocene. The loss of a biotic home, a formative and eternal piece of biotic identity that will become more distinguished as we progress into space, will manifest in our constitution of new environs on Earth or Mars. Even hopelessness in the Anthropocene can be revised in the Astropocene.

In another of Chris Pak's feedback loops sits the very turn from anthropogenic impact to control. As Pak parses "terraforming" and "geoengineering" as terms, he quickly engages Fogg's (1995) notion of making Earth more Earthlike and finds in the etymology a human alienness (Pak 2016, 1–3) that "reflects a shift in awareness of humankind's ability through climate change and other global effects. . . . This sense of terraforming as an extension of anthropogenic climate change illustrates a connection between climate change and geoengineering and by a further conceptual extension geoengineering and terraforming" (Pak 2016, 3). In this appears an odd reversal of potency and scale. Terraforming invokes the fantastical elements of SF—we need look no further than the quintessential red weed terraforming Earth in *War of the Worlds*. "Terraforming" categorically contains the human shaping of exoplanets and alien shaping of Earth as well as the human shaping of Earth by geoengineering.

Geoengineering is but a subset, specifically reacting to earthly climate change. The difference between geoengineering and other terraforming is a product of human alienness, the strange doubling, that occurs from living in the Anthropocene. Geoengineering has shed its fantastic qualities. High-concept sci-fi technosolutions—seeding a planet with modified organisms, injecting chemicals into the atmosphere, or constructing a solar radiation shield in space—are increasingly "on the table" technologically and economically, if not yet politically. Additionally, Earth's climate crisis initiates a clock for our species that is moving far faster than our ability to terraform and colonize another planet. The revelation of the anthroposphere combined with the ability to manipulate the form of our impact initiates the late-stage Anthropocene.

Unbound from the need to know which geoengineering feat we deploy (they are all of the same character), we can interrogate the late-stage Anthropocene. There seems no intention in human action to revert to a more primordial planetary ecology. Disseminated by the SF terraforming fantasy, Pak's characterization of the human-alien conjures an unimpeachable, almost casual, control over planetary ecology where "a compromise between the alien and indigenous is reached" (Pak 2016, 12). In those narratives, our species is at stake. What happens as we fill that alien-terraformer role? What species will we deem unnecessary? What will we do to technologically replace a species that we need but cannot survive terraforming? What will the compromise look like? The answer is very human. We keep or construct what we need to survive. We take control. Terraforming is the sort of thing we always do, arriving again at the Promethean fire-grab with the first epigraph of Pak's terraforming tome: "Making your world more habitable began on the Earth itself, with the first dancing fire that warmed its builder's cave" (Reed 2001, 199; Pak 2016, 3). Indeed, "Terraforming is an ancient profession" (Reed 2001, 199). Geoengineering seems less about making Earth like itself than making it for us. We are not looking for Yellowstone to seep from its borders, continually disregarding that elk seasonally do (Cheshire and Uberti 2016, 66–69) and legislating as if wolves do not. Oddly, this makes terraforming another planet, or post-apocalyptic Earth, more likely to replicate an organic Earth than the late-stage Anthropocene.

Until now, I have used "simulate" or "approximate" to describe all extra-planetary iterations of earthly environs. The Hab approximates Earth environs while Watney simulates an ecosystem in microcosm. Exactness was never the intention. Nor is it intended by geoengineering, which reinstates a planetary average and is not necessarily concerned with biotic interconnections that traditionally maintain equilibrium. In "replicate," something else occurs. The *Oxford Dictionary* has the verb, "to replicate," describing the making of an exact copy while the noun, "the replicate," is defined as "close or exact."[3] The action is to copy perfectly, but there's some underlying imperfection once the replication is complete. Seemingly, even a perfect reproduction is not the original, and even an indistinguishable copy is missing some core authenticity. This impossibility of the replicate to replicate is embraced by the terraforming processes of *Seveneves* and allows a simultaneous invocation of natural process and the absolute inability to produce the former planet. Humanity approaches being nature by releasing control over final form, instead setting up a system that functions as nature.

The mediated process by which the animalia is introduced to a biome shows the limit of humanity. Humans cannot replace nature or oppose its processes. They do not try to recreate "simulacra of Old Earth biomes . . . it would fail anyway because the forces of natural selection were unpredictable and uncontrollable" (Stephenson 2015, 597–598). While they may be able to initiate the entire planetary system and dictate some of the spheres, humans cannot recreate the chaotically arrived at contours of the previous Earth system. If they try, it will fail. Natural selection will not be perpetually directed, not with a resulting self-sustaining balance anyway. The results cannot be chosen, as though natural selection functions in such a manner. They are beyond returning to the prior system so just use its process to maintain the resulting environment. Instead, humanity relegates Old Earth to sentimentality, which is both true and rhetorically persuasive.

Terraforming and the Astropocene are characterized by humanity's ecological mastery. We do not master nature, but master all that can be controlled and leave the rest to sort itself by natural process. Instead of sequencing the genome of a novel subspecies of canid that would successfully integrate with the biome and effectively breed, they "produce a race of canines that would, over the course of a few generations, become coyotes or wolves or dogs—or something that didn't fit into any of those categories—depending on what happened to work best. They would all start with similar genetic code, but different parts of it would end up being expressed or suppressed depending on circumstances. And no particular effort would be made by humans to choose and plan those outcomes" (Stephenson 2015, 600–601). Humans are able to immensely impact the planet. The Astropocene is constituted by a closer relation to the biotic processes of the universal ecosystem rather than the product.

REVERSING NATURAL SELECTION

The biotic identity that doubles humans in the Anthropocene and ecometonymically describes the pre-Anthropocene environs all funnels toward terraforming (felt in our present as geoengineering). This is the culminating productive technology of the environmental sciences, demanding and signaling the intricate understanding of the ecological mesh of the universal ecosystem and sustainable human relations within. The Astropocene is not an ejection from nature by human ecological control but rather re-interrelation to live in response to and maintenance of the ecosystem at its planetary and universal scales. The epoch is characterized by relation to the universal ecosystem, becoming cyborgs within that environment, constructing genomic hyperobjects, and whatever else manifests to allow the simultaneous gathering of maximum ecological control and invigoration of natural, self-sustaining processes. Everything humans have learned approaches the astropic singularity of terraforming. In the Astropocene, humans reverse natural selection so that the environment is instead selected by our physiology.[4]

In the wilds of New Earth, Kath Two is "startled awake by patches of orange-pink light cavorting across the taut fabric above her. A very old instinct, born on the savannahs of Old Earth, read it as danger: the flitting shadows, perhaps, of predators circling her tent. During the five thousand years of the Hard Rain, that instinct had laid dormant and useless" (Stephenson 2015, 569). Waking on the second iteration of Earth, Kath Two displays a formative biology that maintains pre-Anthropocene survival responses, even after the species spends millennia aboard a space station. The environmental stimuli affirm her biotic integration with the terraformed planet—that ecological link between an organism and environment. In Kath

Two's biotic relation to New Earth, *Seveneves* describes a new human relation to nature for the Astropocene—novel relation to natural selection impossible in the Anthropocene.

Kath Two recalls that, tectonic plates aside, all features of the planet have been orchestrated by humans. The molecular composition of the planet's water and atmosphere were imported and designed. All organisms, except humans and the beings constituting their microbiome, went extinct and any iteration exists only as a human decision to reconstitute them from their DNA. Some were bioengineered to make themselves extinct after their role in the terraforming process was complete. Still, the atmospheric composition triggers something innate in humans: "It made people want to be on Earth more than anything" (Stephenson 2015, 656). This truth runs opposite every attempt to simulate a planetary ecosystem on the space station. They are mere simulations. Visiting New Earth, Kath Two is biotically entwined with the environment, responding physically and emotionally, despite rational understanding that it is a replicate of Old Earth. Nothing undermines her biotic relation to this environment.

Activating a survival response while sleeping, though, is a progression beyond the biotic relation to the planet conjured by breathing in the atmosphere. The world has recently been seeded with apex predators. *Seveneves* describes the process of (re)introducing apex predators to the ecosystem with the broad category of canids, offering that they may become wolves, coyotes, or some novel variant, whatever the environment selects. Mentioning wolves as critical ecological components, *Seveneves* joins Yellowstone and McCarthy in describing the reinvigoration and stabilization of the environment by apex predators. And how could Stephenson not also choose the species most likely to congregate in packs and families that mirror and entwine with human civilization? With apex predators, New Earth is wild. In such environs, Kath Two finds human survival instincts remain intact.

Subtler than the physiological impact and connectivity with genetic history produced by breathing the replicated atmosphere, Kath Two also interacts with New Earth visually. About to return to the space station, Kath Two ruminates on the biotic relation between her physiology and the environment. She considers how human brains "evolved to live in environments like this, and to be nourished by such complex stimuli" (Stephenson 2015, 572). She gazes out across the wild environs, looking over a sprawling plain to a lake and across to the tree line. She physiologically responds in the vein of the picturesque. The planetary environment is bodily proven to be preferable to the space station. She describes the innately human desire to "see beauty and purpose in the 'natural' world" (Stephenson 2015, 572). *Seveneves* issues an essential appreciation of the landscape built into humanity by the very information gathered from the environment. The planetary environs are comprised of immense fractal variance, individually accessible and coalescing into enormity. The space station habitats are no such thing. Appreciation of the natural scene's nuance is tied up with the survival necessity of deciphering it. Terraforming, though, makes murky the delineations of such appreciation. Kath Two experiences an aesthetic appreciation with a brain evolved by such a wilderness. Biotic interaction with the scene is somehow linked to aesthetic interaction.

Once again, the Astropocene prioritizes biotic relation. The other visual stimulus Kath Two experiences is the shifting light that wakes her. Registering through her eyelids, recognition of the irregular patterning instinctively invokes fear and alertness, all while sleeping. In this moment, Kath Two is a prey animal reacting to stimuli. This is another sort of nourishment. Recalling the wolves harrying antelope in *The Crossing*, the prey animal is the one invigorated as though by an internal fire. An unexpected seat of primordial fire, the prey animal is as invigorated and sustained by its predation as the apex predator is in its hunting. In the wilds of New Earth, "one did not pee without looking around first, did not venture into the open without a weapon in hand" (Stephenson 2015, 571). The pastoral disappears, or was never produced, as open areas did not facilitate an extended pastoral pause. Gazing at the landscape is a survival technique long before it becomes a luxury. Biotically entwined, Kath Two chooses an open space with unimpeded, long lines of sight. Biotic existence within predator-prey relations entrenched in pre-landscaped and uninhibited wilds—largely a nonissue in the Anthropocene—is critical to gathering ecological control while maintaining nature.

Terraforming reverses the initiating circumstances of natural selection, instead selecting a planetary environment to suit a biotic identity. The planetary environment normally comes first but terraforming in *Seveneves* reissues the current natural state of the planet as an ecosystem that suits human anatomy. I do not wish to unequivocally state that environments exist a priori or as a sort of first cause, but the thought serves. The physical contours of the universal ecosystem—planets, stars, nebulae, black holes—do not evolve so much as settle or form. There is no requirement that life exist anywhere in the universe for these structures to take shape. The reverse is very much true: the universe must have certain structures for life to exist. Reversing natural selection only necessarily occurs in the initiating terraforming action. Relinquishing such ecological control in favor of letting the new Earth system maintain itself is hardly innate to any terraforming action. Terraforming pivots on the human ecoethical framework initiating it.

The first act of terraforming will be to suit human bodies, responding to being doubled on a planet. There is some definitional work to be done when we start discussing terraforming Earth without a cataclysm but no species will be more suited to a terraformed or geoengineered planet than humans.

Seveneves is right: there is no replication. Terraforming cannot reinstate a prior iteration. There are too many variables. Existence is too chaotic. Natural selection will occur by the existence of organisms in an environment. Humans will never control natural selection. We can control the atmosphere, probably enough of the hydrosphere, and we can manipulate the biosphere and everything else. We cannot control the biotic process of natural selection. Nor do we exit it. *Seveneves* offers a vision of how to control everything necessary to live well on and with a terraformed planet.

While some abiotic factors are malleable, gravity will not be pressed into service, at least not sooner than terraforming. *Seveneves* does not have to worry about how natural selection unfolds in such environments. I suspect we will find out. As we spread into the universe, we become very earthly beings. Our biotic identity will dictate our existence for a long time. We will continue to learn that nature is an unimpeachable supersystem that we cannot extricate ourselves from. We can reverse it, so that an environment is selected by our physiology instead of the environment dictating an organism's viability. Terraforming funnels everything down to that singularity. Once the terraforming takes place though, it rapidly propagates outward by the direction of natural selection. Humans are included in that biotic relation because, well, we always were. Nature is biotic interrelation, is natural selection. This is the pinnacle of the Astropocene: being able to construct a natural, planetwide ecosystem that will suit the needs of our singular species—reversing natural selection, so that the environment is, instead, a product of human physiology. The ecoethic appears when humans decide how to handle their relations to the wilds that sustain them. Will humans reciprocate the maintenance by biotic interrelation, or institute scheduled or perpetual maintenance? What does any of this mean for being human in the Astropocene?

PROPERLY SUZERAINS OR SPEAKING AS NATURE

Reversing natural selection is a forceful escalation of control over planetary ecology. Directly inherited from the late-stage Anthropocene's shift from impact to control, terraforming signals ecological mastery. Humanity is still far enough from terraforming a planet that a guiding ethos is not yet required. However, terraforming is a sort of crest that our entire ecological perspective moves toward and what shakes out after is inextricable from what we do now. Creating the reciprocating ecocritical action escalates the import beyond the normal feedback loop of SF and literature, moving from an amalgamation of the future from SF past to a firm future where the present will determine not the action but the quality of action. SF terraforming is beyond informing on present environmental relations by allegorical circulation. The relation to nature we cultivate now determines the contours of terraforming.

Humans in *Seveneves* are lucky. They are absolved of finding another planet for their first terraforming. The tectonic intersection of geology and geography remains quite similar to Old Earth (Stephenson 2015, 644–645). Add the right amount of water to that lithosphere and much of the planetary system can unfold as Old Earth did. Then, control the composition of the atmosphere and enough of the hydrosphere, and even the weather likely accords with the Old Earth models. Humans in *Seveneves* have not experienced weather in 5,000 years, but I suspect everything shy of the most extreme iterations is balanced by having survived in the vacuous environs of space for that time. Earth's orbit and tilt is unchanged so related climate and seasonal effects will proceed roughly as before. Humans have been preparing for their return, so Earth gravity is no surprise to their bodies. Incorporate plantae and animalia and the biosphere will proceed along understandable lines, if not truly precedented. The findings of biology are unlikely to be radically altered by the new terrans who are modified but based on Old Earth structures and categories.

In terraforming, the environmental sciences gain a powerful and novel product. The natural sciences—biology, chemistry, physics, geology, meteorology, and their like—all begin to funnel into the terraforming action of the environmental sciences as it moves from studying or mediating environments to producing them on a planetary scale. Engineering and bioengineering will play critical roles in the terraforming process. All disciplines will shift, overlap, morph, and redistribute for terraforming and the Astropocene. Many such fields have recently emerged as astro- and theoretical physics are gathering paradigmatic information about other planets in the universal ecosystem. All of this is circulating, creating, and being recreated, with self-sustaining earthly ecosystems formed elsewhere. The universal ecosystem offers a new concept of ecological interrelation and terraforming questions fundamental concepts of being a terran in nature. Off-world terrans initiating a planetary ecosystem according to their own anatomy resonate with a version of speaking as nature. By imaging humans relating to nature, ecocriticism provides models of humanness that connect the Anthropocene and the Astropocene by an ethic epitomized by speaking as nature.

If ecocriticism and the environmental humanities fail, the characteristic environmental control of the Anthropocene propagates the Astropocene, and humans terraform as would the judge. The judge speaks of the observation axiom of natural science and stakes his claim on the very world through such parsing of life: "that man who sets himself the task of singling out the thread of order from the tapestry will by the decision alone have taken charge of the world and it is only by such taking charge that he will effect a way to dictate the terms of his own fate" (McCarthy 1985, 208). In the judge's mouth, scientific study is a malicious invocation, but it does not take much for him to make it so. Look around. We have used science and technology to secure our perpetuation. The judge is correct. Scientific knowledge of the natural world allows us to dictate our fate. He explains, "Only nature can enslave man and only when the existence of each last entity is routed out and made to stand before him will he be properly suzerain of the Earth. . . . A keeper or overlord. . . . A suzerain is a ruler even where there are other rulers. His authority countermands local judgements" (McCarthy 1985, 207). Crafting his authority by assembling science to his will is easy. The judge would very much enjoy reversing natural selection to ecologically engineer the planet as determined by the human body.

Terraforming as a human relation to the environment is also a site of challenge to the judge's fundamental anthropocentric control. The judge conceptualizes "suzerain" by forging a synonymity between keeper and overlord. Amending keeper with overlord, the judge disintegrates from the former any implication that a keeper might maintain a balanced natural system. Instead, the judge asserts that a suzerain's authority is the inorganic system instituted and is maintained by its lord. Blending these terms imbues a suzerain with a power derived from actively dissolving any other superseding authority. The judge imposes systemic direction, snuffing out that which he does not consent to and altering things to keep, maintain and perpetuate, what he lords over. This action puts the judge in direct conflict with natural selection, as it is the process by which nature includes humans. Or, as the judge: natural selection is the process by which nature establishes its suzerainty, undermining any human authority. The judge already underwrites all authority manifesting anthropos and recognizes that nature is the only supersystem, the only authority, beyond humans. They are contained within it and natural selection can countermand the authority of the judge. The judge would not submit to the unbound influence of nature on New Earth. Natural selection is too pervasive and imperceptible, too wild.

Elsewhere, I describe the reduction of nature by a planetwide city of such scale that the surface of the planet is lost to its residents (Gormley 2019). A heroic order, once deriving power by aligning with nature and enacting its will, is now defined by a fallible nobility and wrapped up in political machinations and war—all of which is unfolding according to the desires of a single enemy, the culmination of an ancient line. The natural order became a noble order, easily destroyed by a non-biotic politic. Those who might speak as nature, speak against anthropos, are destroyed. Correctively, the order's totemic revenants disappear into exile, entrenching themselves in the wilds of a desert and a swamp, reintegrating with uninhibited wilderness to await their chance to facilitate the hero of the next age. I suspect the judge would also make the planet a city but lacks a method of sufficiently extricating humanity from the landscape. Instead, he proactively settles the desert of the American southwest, delivering the Anthropocene to those wilds. His suzerain sermon concludes with the assertion that he would have all birds in a hell-like zoo because their flight and thus freedom undermines his suzerain authority. This becomes *Blood Meridian*'s epilogue, which becomes *The Crossing*, which becomes the present, then *The Martian* and *Seveneves*. The judge would speak as nature by containing and replacing, by permeating everything with anthropos, by not permitting anything to unfold without his consent. The judge would speak as nature by silencing it and ruling in its stead.

Opposing the judge's zoo, *Seveneves* describes New Earth as an ecological preserve (Stephenson 2015, 597). A preserve implies human keepers or wardens promoting, if not always perfectly maintaining, wilderness unfolding as itself, an ecological mediation of anthropos and environment, not unlike the wolfen otherworld attempted in Yellowstone. In terraforming, the preserve's borders become the planet, extending to the edge of the atmosphere or at least the biosphere. Such totality makes opposing the judge a bit simpler as terraforming to suit humanity implies the intent to live biotically uninhibited on a verdant planet. One must be entrenched in biotic relation on a terraformed planet. The suzerain as individual implies time spent in a location exercising authority.[5] The judge is constantly exerting his authority over the leaders in his environs, deciding and influencing without leading. He is an authority without office, a difficult position to maintain in the wilds. We know that apex predators are critical components to self-sustaining ecosystems, and it is far easier for humans to put them into the ecosystem than to be them.[6] Immediately, humans become a prey species option. Kath Two describes that while the ecosystem should balance itself, she cannot be certain that a predator is procuring enough food or would not see her as an opportunity to expand its resources (Stephenson 2015, 571). Further, what microorganisms or viruses may have evolved unknowingly? Could plant spores induce an allergic reaction and anaphylaxis, or some darker SF or ecohorror trope? Being within terraformed nature immediately creates biotic relations described by natural selection.

As a genetic extension of the Anthropocene becoming the Astropocene, Kath Two is a unique approach to speaking as nature by reducing the epochal swaths necessary for a biotic identity to shift in the landscape. As the humans of *Seveneves* settle on a moon fragment, they find that there remain only seven women with reproductive ability. Their geneticist explains that parthenogenesis, a process creating an embryo from just an egg, is as simple and occasionally easier than the repairs she would need to do on the available sperm which is likely damaged from exposure to cosmic radiation and the station's reactor (Stephenson 2015, 553). Additionally, these last Old Earth,

Anthropocene terrans agree that they will each get to select one trait to genetically code into their offspring—intelligence, aggression, or whatever else is
known to be expressed in the genome. Not always so reductionist, one Eve
angles toward chivalric heroes while another chooses a different trait for each
of her individual children. These selections are made based on a self-reflective value of characteristics contextualized mostly by continuing the species
in space habitats for thousands of years and occasionally responding to each
other's choices, sometimes in dramatic and spiteful ways. Understandably,
there is no talk of the value of these traits in earthly environs and each has
varying degrees of value outside the spacecraft.[7] Exercising the most foresight, Kath Two's Eve was Moira, the geneticist. Moira chose to hyperac-
centuate epigenetic expression.

Epigenetics describes changes in gene expression without altering the
nucleotide sequence. Molecules attached to DNA alter the level of activity
of that particular gene, regulating the expression of that trait. The molecules
might randomly detach, but often "epigenetic changes occur in response to
our environment, the food we eat, the pollutants to which we are exposed,
even our social interactions. Epigenetic processes occur at the interface of
our environment and our genes" (Francis 2012, xi). Moirans are designed to
be extremely physiologically responsive to these shifts that would normally
manifest mildly in others. Colloquialized as "going epi," Moirans' epigenetic shifts result in massive swings in character by the novel expression
of their genes. Kath Two gets her nomenclative number as she has shifted
once before. While the other Eves selected adaptations that seem mostly in
response to what they foresaw as necessary to ensure the species, Moira chose
to let the environment select what was valuable. Normally, natural selection
requires generations to encode into a species the biotic identity balanced with
the environment. Responding to potent stimuli, Moirans can adapt to their
environs in hours.

Kath Two becomes Kathree. Tyuratam Lake, as one genetically predisposed
or fated to do heroic things, believes that Kath Two goes epi in response to
seeing her longtime mentor and his assistant killed in the botched first contact
with the Diggers, that branch of humanity who survived the Hard Rain underground in mines. Such a trauma can cause an epigenetic shift in Moirans, but
Kath Two shifts in response to combat instead (Stephenson 2015, 837–838).
The "chivalric hero" assumes that it was the traumatic viewing, but Kathree
emerges from shooting a Digger in the stomach who had just struck her
arm with a shovel. Kathree identifies that the shift occurred in the entangled
duality of being attacked and fighting back. Going epi, while not altering the
genes themselves alters the organism's phenotype, the observable expression
of the genotype. Afterward, Kathree vomits from the sudden hormonal shifts
and is at risk of testosterone poisoning. A medic gives her probiotics to settle

the resulting disruption of her microbiome as she is "colonized by any old germs, including ones from the Diggers that had never been exposed to a Moiran body" (Stephenson 2015, 788). Especially indicative of biotic relation, her microbiome's disarray will ultimately resolve, tying Kathree's biotic identity to the critical kin of her immediate environment. Her microbiome becomes a balanced system ecometonymically referencing the microorganisms inherited from Kath Two's gut and skin, those roaming around on New Earth, and the bacteria and viruses composing the microbiomes of Kathree's immediate social connections. Moirans generally consume a lot of food after going epi, which also contours the microbiome. Emerging a completely new organism, Kathree is now a phenotype expressed in response to the current environmental contours that her prior biotic identity was ill-equipped to handle. Her biotic self has evolved to suit the immediate environment.[8]

The epigenetic shift secures successful biotic relation by immediate natural selection. Collapsing the timescale of evolution-like process by increasing the intensity of epigenetic expression of phenotype makes Moirans more equipped to maneuver the return to natural environs after millennia aboard a spacecraft. Biotically responding to the combat circumstance of her new environment, Kathree is now significantly faster than the most physical of her companions and has an enhanced sensory perception, including elevated hearing and smell. Now Kathree can "sense . . . footfalls through the ground" (Stephenson 2015, 812). This ability is always distinct from hearing footsteps, though she can do that, too. The sensing is also kept distinct from a strictly tactile experience. It is never described as feeling but maintained as the categorical sensing, issuing either a combinatorial, holistic sensory experience or that the ground is conveying some novel sensory experience beyond traditional human senses. In either case, Kathree is so attuned to the environment that her sensory awareness now includes those moving within the immediate landscape. The potent immediacy of sensing footfalls through the ground is reminiscent of tracking, although Kathree, as with natural selection, collapses the temporality of the tracking action.

The very act of tracking implies that the organism's present location is unknown, with the usual intent being to pursue quickly enough to collapse the temporality into a simultaneous present shared by pursuer and quarry, especially in the violent biotic context from which Kathree emerges. Sensing footfalls is to recognize the organism in its environment as it makes the track. She can sense the information of a footfall as it registers the biotic information in the ground and forms the track. Kathree can sense organisms through the ground as far off as the very edge of her sight. Further elevating her suitability to a violent environment, Kathree can sense tracks being left in a trackline she cannot witness, such as one being made by someone pursuing her who is otherwise silent and more adept at combat. Kathree's epigenetic plasticity always responds to her immediate circumstances, radically altering genetic expression and creating a superior biotic self in conversation with the environment.

Moirans approach personally employing natural selection by dramatically deploying epigenetic expression of phenotype. Instead of minimizing or mitigating relationship to the environment, Moira leans into the biotic inter-action and makes a race of humans who are designed to be entrenched in their immediate relationship to their environment. Inheriting this malleable biotic identity, Kathree stands opposite the judge. Instead of mitigating wildness and mediating biotic relationship to the environment, Kathree is enmeshed. Moirans are not bound to the arduous process of encoding a successful biotic identity in DNA. Rather, they alter how their genetic material is expressed, modeling the process of natural selection. Going epi is only a model as the actual genetic data remains the same, but the ability to manifest available genetic material enables one to be their most capable iteration. Managing the transition symptoms seems to be their most innate social drive, although the microbiome is also a construction of socialization—an enhancement from Watney's food production depending on isolation. Instead of minimizing and mitigating biotic interrelation, Kathree is intimately bound to her immediate environs and other organisms.

If ecocriticism and the environmental humanities prevail, we get to live unlike the judge as Kathree does. While invoking the process, Moiran epigenetic shifts do not actually encroach upon the function of natural selection. Altering phenotype without changing genotype, going epi only simulates an organism manifesting powerful survival traits. All information must already be in the genome as the only alteration is the degree to which it is expressed. Further, going epi is completely biotic. The body responds to environmental stimuli that cannot be handled by the current phenotype and alters that expression. Moirans do not choose to go epi. Nature still selects. They simply have a greater range and more dramatic reaction to the circumstances that already cause epigenetic expressions. Kathree's hyper-reactive biotic identity models an ecoethical framework antithetical to the judge's.

Kathree's epigenetic shift bolsters the archetypal ideal of one who speaks as nature. Biotic relationship to the environment, especially wild nature, determines what characteristics are successful. Kathree is no more doubled than us, Anthropocene or Astropocene. The difference is that her mechanism for responding is biotic integration, to become an ecometonymic expression of the immediate environment. She is freed from eternally referencing the inorganic, manufactured habitats of the space station. Kathree responds to the distress of incompatibility with immediate circumstances through biotic plasticity that internalizes the external environmental features and the internal and external organisms of other's microbiomes as she reiterates her external phenotype and microbiome. Just as nature, Kathree cannot be described by the internal/external binary. Moirans are inoculated against becoming doubles in their own environs, insulating them and their offspring against our doubling in the environs of the Anthropocene. While the wild environs are indeed terraformed to suit human physiology, they are organic and mark the ability and, more importantly, the desire to construct a balanced Earth system.

Ultimately, the reason we cannot speak as nature is that we do not exert its processes, only exemplify natural selection as ecometonymies expressing and expressed by nature. The judge would subvert natural selection, conjuring a system in which his authority countermands the process. He would harness nature and make eternal that first terraforming process by which human anatomy selects nature. Kathree models a hyperinteractivity with the process by which nature unfolds into more nature. Alternative the judge's ecoethic, a Moiran's biotic identity responds to the environment itself for reciprocal biotic interrelation. While Kathree is stabilizing, canids are constantly howling, also going epi. People fear that Moiran proximity to nonhuman species going epi would inspire epigenetic shifts that entangle Moirans and animals in some perceived negative way—I suspect by its wildness. This is never actualized in *Seveneves*, allowing us plenty of room to read that such distinctions between human, animal, and environment are fabricated noncategories because evolutionary processes always occur by the interactions of species. The locative signaling of these wolfen howls instead parallels the ecometonymy of Kathree's new biotic identity determined by interrelation with her environment.[9]

Kathree emerges violently capable, although never malicious, and becomes a sort of ambassador to a third, ocean-dwelling iteration of humans who survived the Hard Rain by selective breeding. Prioritizing biotic identity conjures an ecoethical approach to terraforming where, after the terraforming reversal, humans reinsert themselves into the processes of natural selection for the maintenance of a restored planetary ecology. I issue with such fervor the idea that we might speak as nature because I can only hold off for so long the recirculation with the present inherent in all SF. Before moving forward with the Astropocene, there are some lessons *Seveneves* imparts. While those humans did not have to suffer the Anthropocene, they did fail to resolve some of its most harmful intersections. As long as there is no cataclysm, we may be able to do enough work so that our destructive histories do not dominate the Astropocene.

WHAT IS HUMAN IN THE ASTROPOCENE?

The Astropocene epoch marks terrans ecometonymically describing Earth in extraplanetary environs, thus actively connecting biotic Earth with the universal ecosystem in novel ways beyond things arriving randomly instead of just passing by or by the gravitational and photonic contours of the universal ecosystem. Beyond even the technology-based shifts to the universal ecosystem that occur in the Astropocene, organic systems are impacted and curated in novel ways. At the pinnacle, terraforming constructs environments with nothing more than abiotic planetary characteristics and technological control. The Astropocene is characterized by the biotic interaction of terran humans with the universal ecosystem. Humans in the Astropocene are the same basic stuff as ever: an amalgamation of biotic and ontological identity. The difference is that in the Astropocene, biotic identity is as foregrounded as ontological construction of self has been in the Anthropocene.

Aligning with biotic and ontological identity, ecocriticism approaches the future via two analytical frames, the SF feedback loop and the reciprocating ecocritical method. Applying both to the latter arc of *Seveneves*, selected for its dedication to ecorealistic interactions of character and setting, illuminates the role of ecocriticism in describing and shaping the Astropocene. I would draw attention to three events in the latter arc of *Seveneves*—establishing survival methods for the species, staking a claim on New Earth, and the novel's falling action—for triangulating the biotic intersection of human and environment in the Astropocene as assessed by the two ecocritical approaches. The remaining humans, later called Spacers, make survival decisions at the end of Part Two that directly manifest in the characteristics of their offspring 5,000 years later where Part Three of *Seveneves* picks up the story. Unseen, the mine dwelling proto-Diggers and submarining proto-Pingers are each making their own preparations.

As the epoch closes, three distinct groups of humans are all making decisions about how to secure the species in extremely restrictive environments for an unprecedented length of time. The future biotic identity of the Spacers is established through genetic engineering. The Pingers deploy selective breeding to control the manifestation of traits beneficial to oceanic life, while the Diggers respond to their limited space with a breeding program and sterilization. In the final climactic event of *Seveneves*, these groups lay claim to the surfaces of New Earth resulting in conflicts and alliances that manifest as the tensions of Old Earth or the Anthropocene, whichever you prefer. The falling action of the book settles these first contact conflicts and considers the Purpose, a secret society-style concept that acts as the novel's call to action as contoured by the group dynamics of the main characters. The SF feedback loop tends to favor ontological concepts of the posthuman, transhuman, postcolonial, and capitalism whereas biotic relations for ecoethically reiterating ontology feature when the reciprocating ecocritical method is applied.

The Astropocene with the SF Feedback Loop

A posthuman lens zeroes in on the decentralized notion of humanness, which *Seveneves* does well across many levels. The novel dismantles humanity into three subexpressions of the species, each with their own social delineations. Spacers consciously create unique races while tensions brought to a head by terraforming split the seven into two factions. The Diggers seem to create a totalitarian class system. Without the finite space of the mines, Diggers quickly invoke personal agency in response to Spacer politics. The Pingers are least explored although, visually and phenotypically, are the most ecometonymically expressive of their environment. Oddly, the three groups do not seem any more posthuman than humans in the Anthropocene, everyone a casual cyborg. Spacers have mastered life support and genetic engineering. The Pingers are metallurgists and gatherers, sustained by resources scattered on the ocean floor. These people are naked, except for net-like harnesses used to carry their technology, the amount of which is suspected to indicate certain roles. The Diggers appear least futuristic but are no less entwined in technological abilities. They manage to preserve the wooden handle of a Craftsmen shovel for 5,000 years. Broken and discarded, it is a material sign revealing the Diggers to the Spacers as another group of humans predating the Hard Rain. Epitomizing the inheritance of our (post)humanness, this artifact from the Anthropocene initiates a cultural clash derived from perceived insults and is then used to kill two people—likely to be cited as the event catalyzing the war for New Earth. The posthuman reading arrives at the unenthusiastic conclusion that we are the humans of *Seveneves* with or without their technology and its application.

The prior reading of Kathree as biotically integrated is most aligned with ecocritical posthuman perspectives. The Spacers are worried about Moirans being near animals who are going epi, fearing a reciprocity that would decentralize humans among animals, as though they could become weirdly indecipherable. This is never delivered upon though, and Kathree remains a posthuman whose organic body is inextricable from the bioengineering technology that defines it, although the technology is spatiotemporally quite distant from her body. The rift in the feedback loop is that we are already posthuman and are not seemingly any closer to using bioengineering to be better defined by the environment. Instead, the idea is to use it as they did terraforming, to remove those things that would harm us. The Spacers did this, too, but the extinction-level cataclysm forces a paradigm shift that produces Kathree. With only SF to posit such things, why would *Seveneves* be more likely to produce the feedback loop than anything else? The productivity of the SF feedback loop in the Anthropocene is especially problematic in a global society that largely privileges scientific knowledge and ability to produce Kathree over the cultural and biotic application to humanity.

The transhuman perspective gains a bit more traction with the differentiations among terrans in *Seveneves*, especially as physiological and cultural distinctions react to environment. For the Spacers, parthenogenesis creates embryos. Further modifications remove known genetic defects, physical and psychological, while also inserting or accentuating characteristics chosen by the parent. The Pingers deploy selective breeding to respond to their oceanic environs, eventually able to survive underwater by their physiology alone. The Diggers, most concerned with population outstripping available space and resources, establish a breeding program and sterilize members of the population, creating their culture. The Pinger phenotype is the most visually dramatic shift, as the group manifested blubber that insulates and streamlines, skin flaps to protect genitalia, sharp teeth for eating fish, and a skin tone for underwater camouflage (Stephenson 2015, 852). The Diggers tend to manifest the red or blonde hair so prominent in the family who prepared the mine shelter and show a whitening of the eyes consistent with dark-dwelling organisms (Stephenson 2015, 752). Spacers show similar fervent phenotypical perpetuation, though they seem to have more initiating archetypes. The Spacers appear to have more variance than the Pingers or Diggers, although this may be a form of anthropocentrism conjured by the narrative following them.

Transhumanism gathers myriad exciting visions and attaches them to the Anthropocene by our technology and the relative immediacy of 5,000 years. *Seveneves* stunningly remains hard science fiction and proactive use of the SF feedback loop would reduce the time frame significantly. We would not need to make do with the limitations of one expert in a single lab on a space station. The transhuman production in *Seveneves* just creates a bunch of doubles who fight over rights to an environment they are doubled in. Outside of Kathree, none are progressed by the ecology of New Earth. They are fighting for control of a biotic place so as to not be doubled. Instead of integrating though, they create a place based on their bodies, so as not to be doubled at all. This is terrifyingly similar to the Anthropocene, especially through concepts of the climate crisis as threat multiplier and the lack of regulation over genetic engineering (Mach et al. 2019; Trivedi 2011).[10]

Both the posthuman and transhuman in *Seveneves* indicate lingering, postcolonial problems unresolved as the species tries to survive an extinction-level event. The most dissociated of the Seven Eves calls out that the selected genetic modifications for the Spacers just create races. In agreeing to the bioengineering, Aïda pronounces a curse, not on the rest of the Eves, but on her own people. She knows that she does not create the curse, but rather the unanimity of the others silences her dissent. She knows that humanity will never merge into a single human race again, as they are now. Instead, her offspring will suffer the prejudice and she will select adaptations for them so they will be able to bear it. The youngest of them, Aïda births many children for whom she decides on separate modifications, often in direct response to the others' genetic identities. Unlike the singular genetic trait among the other races, the Aïdan race is multifaceted for they needed to be robust enough in and of themselves to withstand the othering that Aïda foresees.

Almost 5,000 years later, the Spacers have split into two distinct entities and respective claims to terraformed New Earth exacerbates the fissure, transposing it onto the Diggers and the Pingers in a potent simultaneity of colonizing indigenous peoples and globalized co-opting of less developed nations. An earthly surface ecometonymically described by biotic humans brings the Spacers, Diggers, and Pingers into contact. Even with an unsettled planet met upon by three groups descending from humans, the violence of the SF first contact trope is inescapable. The Diggers and Pingers have already claimed, respectively, the land and sea of New Earth. Stephenson never resolves the lingering postcolonial problems—perhaps because science provides no actionable information for such social events and processes. Perhaps the late-stage Anthropocene is where such a resolution occurs, and the cataclysm of *Seveneves* removes the space for postcolonial repair of human culture.

Worth noting in SF, such a colonizing stance is only viable in this scenario. The Earth system was destroyed, leaving only the barest sliver of terrans, humans, and their microbiomes.[11] The classic SF trope of speciated differences and literal alienation are undermined by the very dictate that these groups are all equally as (post)human as any of us. The potential for a noncolonial human epoch is immediately undermined as the racial split among the Spacers draws in the Diggers and the Pingers and takes a wholly political and militarized form that was built into the Spacer races even with Aïda's forewarning. Once again prioritizing borders and race, everyone is about as far as they can get from a biotic relation to the planet. The SF feedback loop shows how these things are entangled with the environmental humanities and, without resolution, continue to define any iteration of humanity, even after the end of the Anthropocene. Visions of the posthuman and transhuman are subject to disruption by unresolved postcolonial humanity.

Surprisingly, capitalism remains undemonized in *Seveneves*, primarily manifesting as TerReForm, the company that terraforms New Earth, and the Crow's Nest, a bar owned by old money Spacers that is so notably unprofitable that it circulates within the mysterious notion of "the Purpose." Terraforming as a product is a strange concept in *Seveneves* as it makes a planet habitable for a species. The fact that it is a business with such an agenda signals governmental subsidies as well as consumer-level products. I imagine things like selling authentic Old Earth potted plants or pets, and I wonder about TerReForm's rights to the surface itself. The company's name does become synonymous with the terraforming process as described and governed in treaties concerning the recreation of Earth. All of it emerges from a Spacer ideology concerned with getting the Earth system in place as quickly as possible—a predictable notch against TerReForm as it signals a money grab over patience and proper application.[12]

Much is left undescribed, which is refreshing in its way because this is hard SF. Nothing is left unsaid about the technology. Capitalism is less burdensome on the space station than in the Anthropocene. TerReForm skews positive as an idealistic replication of those space exploration companies of the Anthropocene. Another reversal, TerReForm seeks to produce a stable earthly system instead of leaving. Lacking more detail, I would take the opportunity to read TerReForm with a positive big picture skew: a capitalist structure whose unbounded success is entirely dependent on constructing a balanced planetary ecology that is maintained by the biotic relations of the organic components unfolding natural processes—imagine such a capitalism. Capitalism is underwritten by the goal of reinstituting a balanced Earth system. The big picture plan for the Spacers was always to return to Earth (Stephenson 2015, 589). Beyond this dictate, the specific mechanism by which such a capitalism exists is inversely vacuous in structural detail to the technology of *Seveneves*. Imagine the environmental humanities if such an underlying ethos did structure capitalism, instead of the near-axiomatic opposition of capitalism and balanced environs. The closest *Seveneves* gets is that an ecoethic rooted in biotic identity sorts it out. I do not disagree with this process, but I wish we similarly had already arrived at such a system. I suspect it is easier to conceive futuristic technology than green economic praxis.

As a text, *Seveneves* subtly permeates capital with vague, deeper purpose so it becomes less stifling for Spacers than in the Anthropocene. The climactic interaction of Spacers, Diggers, and Pingers on New Earth is bookended by images of human assembly. For ventures of significance, Spacers form a Seven, a group comprised of one member from each race. Before the events on New Earth, the group gathers in the Crow's Nest where Tyuratam Lake is the bartender. Owned by old money, the bar never makes a profit but rather functions as a locus of information that is more valuable to its owners. The whole thing is rather ominous and deeply unsettles a special forces operator dispatched to bring an item to Earth on behalf of the owners. The soldier contentedly withdraws his concerns over the suspicious paradox of the extreme cost of bringing the item and the bar making no financial sense when he surmises that the errand has to do with the Purpose (Stephenson 2015, 811).

Uninitiated but well positioned, Ty supposes that the Purpose describes some larger ethos beyond the political infighting and actual combat of the Seven's time on New Earth (Stephenson 2015, 860). He suspects that the Purpose describes some better theology-esque notion that extends from the Agent and can positively drive the present. The Purpose is unknown to him, but he notes that those who claim to be guided by it act in a better way.

Symbolically punctuating the talk of the Purpose, the novel closes with an expansive reassembly, having replaced the dead members of the original Seven and expanding to include a Digger and a Pinger. The mission is over, so the Seven is disbanded but in its wake is a group that is bound together by something apolitical, even though it may become so. The Purpose is too vague for me to comfortably assert its thesis, and I am probably too inclined to go green with it anyway—although, how could that be bad? The Purpose does sit at the intersection of the destruction of the moon and Earth, terraforming a self-sustaining planet, coalescing all versions of humans despite political perpetuation of differences, and a green(er) capitalism. Whatever the Purpose ends up being, *Seveneves* illuminates the biotic dissonance of the Anthropocene, allowing the SF feedback loop to be deployed ecocritically.

Otherwise strictly political, *Seveneves* is drawn into ecocriticism and the environmental humanities by terraforming. Pak contextualizes terraforming as the SF trope (re)circulating with geoengineering as a feedback loop "that refigures tropes and narratives in response to the needs and desires of society indicative of human relation to global climate" (2016, 100). *Seveneves* surely continues Pak's analytical trajectory by deploying cutting-edge bioengineering as a transformative technology that traces an organism to the environment to affect and maintain natural balance. I have similarly read wolf reintroduction as environmental maintenance, arriving at the same notion about canid apex predators in *Seveneves* as critical agents in environmental permanence at the planetary scale. Such scale dissolves the institution of borders that attempt and often fail to keep wild within, both as the boundaries rarely restrict animal territory and hunting can keep animals from returning to the park. *Seveneves* refines this by issuing a planetary scale wilderness preserve that links the evolutionary process with interactions in the environs that reciprocally maintain the spheres of the Earth system. Although the novel surmounts the complexities of treating our world thusly by ending it so completely, geoengineering Earth to be more like Earth is a fine enough mediator of the anthroposphere, once a proper ethos is established. I see only one undermining interaction with terraforming recirculating in the Anthropocene via the SF feedback loop.

Terraforming, as indicating an ethos and as an action, tragically always already undermines the valuable posthuman decentering. Terraforming's hallmark reversal of natural selection retains biotic anthropocentrism, harnessing wild process by the boundaries of the human body, never truly interrelating with the ecosystem. Nature is contorted as humanity builds into their environment an anthropocenter that abrades against the innate disruption of internal/external inherent in nature. Let us consider reforestation as a carbon neutral action in this terraforming feedback loop. Consider how carbon emissions are balanced without recognition of the biological impact of the initial action, defined instead as global average temperature. Reforestation never admits the potential erasure of a biome's complex ecosystem or any of the lasting damages by the intentional and unintentional products of processing harvested resources. Such actions constitute ontological and political ethos by rhetorically decentering the biotic, never maintaining nature, just a perception of the relation. In the feedback loop, terraforming in *Seveneves* gets bogged down in the Anthropocene and geoengineering maintains its negative capitalism and the irrelevance of the posthuman. The transhuman only reinforces the postcolonial. All the old stuff spills out and humans are doubled again in the environment.

The Reciprocating Ecocritical Method

The SF feedback loop, as ecocritical praxis for the future, is stymied in a sort of Hume's fork. When reflecting the problems of the Anthropocene by transporting them to the future, SF either helps us resolve the present by dissolving images of the future, or it does imagine what is next and the Astropocene is just the Anthropocene in novel environs with cooler toys.[13] The reciprocating ecocritical method instead envisions the future of biotic relation to the universal ecosystem, contouring that future just as the feedback loop does with the present. The method has the luxury to do so because it does not disrupt SF feedback or any other literary and cultural process, refined to utilizing a much smaller and more recent cross-section of SF texts. Literature inherently stacks its non-mutually exclusive readings and the reciprocating ecocritical method is merely another such tool. The reciprocating ecocritical method describes the imperative of biotic identity as a framework separate from and then informing ontological terran constructions of self. This method can deploy the SF feedback loop or just sit next to it, two microcosms of the larger literature-culture circulation. After all, these are just various orreries moving by the same mechanism. We invented all of it, except the natural processes defining biotic circumstance. Those do not function anthropocentrically, even if our recognition of them is decidedly human. Reversing natural selection only requires altering the initiating circumstances, not the following process. The ecometonymic description of Earth ecology in our species is as directing as any other species, and never extricated from them anyway, productively decentering our species. Humans need to embody a biotic identity in a balanced ecosystem, which the SF feedback loop does not readily identify, instead manifesting in the Astropocene the values that initiated the Anthropocene.

Reading Kathree as a case study of the Astropocene presents an organic cyborg; her defining epigenetic shifts are technologically crafted but after 5,000 years are stabilized and propagated in organic human physiology.

Kathree manifests the characteristic mastery of the Astropocene but, as the prior section describes, always responding to the environment itself without human control or input beyond her unchanging base genome. Kathree approaches speaking as nature by this mechanism. In the feedback loop, she represents this as a metaphorical potential meant to teach us how to understand ourselves in nature as contoured by our current iterations. We get bogged down in epigenetic expression in the Anthropocene detailing an ecometonymic relation to the environment via things like the dramatic uptick in asthma and food allergy rates (Peterson 2017, 127; Baldacci et al. 2014; Tang and Mullins 2017). Humanity is genetically expressing the Anthropocene. We need Kathree to be more than a hopeful metaphor that makes our lives a cautionary tale. She needs to be an actual vision of the future rather than just contrastingly demonstrating hubristic relations to the environment. In the feedback loop, *Seveneves* reveals its manifesting connection to the flaws of the Anthropocene, as does all science fiction in refracting the present. This is not a mark against the novel, the genre, or even the feedback loop, only to say that there are additional critical actions that can be taken.

By the reciprocating ecocritical method we get to collapse the present and the future, as Kathree does by reading tracks as they are made. In both cases, we assess only the character, the broader strokes of the age. Kathree is hardly privy to the myriad pressure releases and minute sign that is undoubtedly expressed in the landscape. We instead trace larger biotic relations to create an ecoethic that connects us to Kathree, an action akin to speculative tracking. Then, we can handle the unprecedented questions of the age to come. Our biotic relation to the environment has been ontologically and politically crafted, which only brings us to Kathree if we give up control once we reinstate wild balance by the inherently anthropocentric terraforming.[14] The reciprocating ecocritical method instead reads the whole process and can participate in a nature narrative of Kathree that we must approach as though the future has already happened. The timeline is collapsed for us as for Kathree.

To make it out of the Anthropocene and into the Astropocene demands a reinvigoration of biotic identity, locating oneself by a maintaining relation to the environment and the universal ecosystem. Ultimately, the Anthropocene is a geological age that is described by the Earth system. We need to face our biotic identity to reinsert it into a new ontological construction of self for the late-stage Anthropocene. The Astropocene accentuates our earthly anatomy, challenging us biotically while also putting us in contact with the natural contours of the universal ecosystem. In that schema, biotic identity details the processes by which nature unfolds, which is not always organic, or gathers new meaning to organic—it hardly matters which. Natural selection is but one extension of the spatiotemporal fabric to which humans biotically relate in the universal ecosystem.

NOTES

1. Unfairly, we need to get through some of those generations before we can fly our personal spacecraft somewhere. I think my personal curse is being able to imagine such futures—as well as entrench myself in, bind my reality up with the imagined realities of SFF stories. I doubt I will get to space. Sometimes, I fear that my generation may not even make it to consumer-level virtual reality that could count as living in that future.

2. This move is inspired by the unique perspectives and relations to spacetime of Dr. Manhattan's simultaneous experiencing of every moment in his existence (Moore and Gibbons 1986), the temporalities of Billy Pilgrim and the Tralfamadorians (Vonnegut 1969), Stephen Hawking running time backward in Penrose's theorem (Hawking 1988, 80), and the time dilation of general relativity.

3. This distinction is not made in the OED, even though both definitions of replicate (n.) are in dictionaries published by Oxford University Press. They describe the difference: "The dictionary content in [Oxford Dictionary Online] focuses on current English and includes modern meanings and uses of words. . . . The *OED*, on the other hand, is a historical dictionary and it forms a record of all the core words and meanings in English over more than 1,000 years, from Old English to the present day, and including many obsolete and historical terms" (Oxford University Press). There is some modern thread that implies a copy's imperfection.

4. This demands an ecoethical fervor. *The End of the Anthropocene* is not that project, although I intend it to provide a clock, a level of foresight and direction. The current efforts of the environmental humanities are embroiled in figuring this out. I guarantee you though, we will terraform a planet. All there is left to determine is our ecological role in that novel but earthly environ.

5. This is logistically difficult when narrowly defining suzerain in terms of states. While the judge certainly evinces such political designations, he is an individual being enacting suzerainty.

6. The Glanton gang is the anthropos pack, a disturbing political foil to the biotic wolf packs of Yellowstone.

7. I see a larger discussion of how these decisions continue anthropos, as well as the way they are discussed. While fascinating, it is just outside of the scope of this project. Anthropos will extend beyond the Anthropocene.

8. The medic also treats her with a drug that prevents her brain from lingering on the distressing vision of her friends' deaths. Stephenson is sure to include psychology in epigenetic expression, although it may be less relevant to Kathree's particular circumstance.

9. I like to think of Haraway's purely chthonic beings, humans with animal DNA in them (Haraway 2016, 141).

10. Did anyone see *Jurassic Park*? There are five of them, going on six, and they all say the same thing.

11. For simplicity, I am proceeding as though whatever other terrans survive in the deep earth and ocean are unimpacted. The Spacers terraform Earth to be like Earth after all. This lets us proceed without being distracted by concepts of colonization when it is humans against microorganisms on another planet. Terraforming Mars in

SF often runs up against indigenous organisms, although we probably wouldn't think too hard about it since they're likely microorganisms—a reality reeking of colonialism. Further problematic, the primary discursive action in defense of indigenous microorganisms is that they are valuable to scientific study, not that they are beings. I'm sure you also feel the rhetorical impotence of "microorganisms are beings too" when scientific pursuit is rhetorically available. Although that is unimportant too if the circumstance is that the indigenous organisms or the human species goes extinct. Capitalism usually trumps both, anyway. What fascinating asides SF offers.

12. This all says nothing about the apparent monopoly of TerReForm, although I wonder if such things are more related to the population size and refined specificity of the knowledge bases needed to make the Spacer colony succeed. *Seveneves* does describe monopolistic weapon production, but that is driven by the environmental circumstance of needing projectile weapons that do not puncture the space station hull rather than asphyxiating monopolies of the twentieth century—although I'm not sure how monopolies do not become asphyxiating, or if they are ever not. The projectiles are largely nonlethal nanobots that incapacitate targets and fight other nanobots. A cold war comes out of it but not by the advent of technological mutually assured destruction. Rather simple, hurling a rock at a space habitat breaches the hull and creates the same level of destruction as a large bomb (Stephenson 2015, 677).

13. This is separate from discussions of technology as predicted by SF. There is a clear line between the futuristic tech and our own technology arriving at it because of it. Here I am concerned with biotic relations, the ecocritical assessment of human relation to nature.

14. Here and elsewhere, I use terraforming to mean geoengineering too. Terraforming's totality absolves accounting for the unpredictable nuance of nature after geoengineering, which may effectively institute New Earth anyway. The reciprocating ecocritical method eventually bumps into more speculative SF iterations but must do so carefully to not disable its predictive value.

Chapter 5

Spatiotemporal Nature

The Astropocene marks a paradigmatic interrelation with the universal eco-system. In this epoch, humanity is no longer mandatorily linked to the blue planet on which we evolved. Our extraterrestrial biotic homes integrate the abiotic contours of the universal ecosystem to replicate earthly environs. In *The Martian*, Watney stores potatoes on the surface of the Red Planet to pre-vent them from rotting and adjusts the temperature of his rover by removing insulation designed to barricade him against the Martian climate. The Spacers in *Seveneves* gather icy comet cores to manufacture fuel and other resources, eventually redirecting them to rehydrate the planet's surface for terraforming. The Astropocene gathers novel exigence to the ecometonymic pathways that trace to abiotic environmental features like molecular resources, thermody-namic descriptions of entropy, and gravity. Interrelation with the universal ecosystem dissolves abstract connotations of the dimensional fabric of reality in favor of biotic relation to a spatiotemporal nature, the abiotic landscape of the universal ecosystem. Ecocriticism must extend its scientific literacy to include physics to cultivate an ecoethic of interrelation that both ends the Anthropocene and initiates the Astropocene.

The Anthropocene has already gathered to itself such physical definitions, granting an oddly inorganic sense to an epoch so concerned with biotic rela-tions. Currently, the Anthropocene is a concept, evocative and productive, but not yet an official geological epoch. The short version is that many of the things we associate with the Anthropocene are not neatly expressed by the planet's geological cross-section (Steffen et al. 2015). While "The keepers of geologic time must build their timelines directly from the physical records of Earth history laid down in the rocks by the geological forces that shape our planet" (Ellis 2018, 34), ecocritics can concern themselves with a wider lens for assessing the character of the epoch. This sort of character analysis is even streamlined by the intense scientific scrutiny given the potential geological markers for the Anthropocene—an indicator and boon for use of the reciprocating ecocritical method. From this vantage, ecocriticism assesses the biotic and abiotic contours of anthropogenic environs to characterize the Anthropocene and its end.

Tracing the biotic expressions of anthropogenic impact reveals the Anthropocene is characterized by dissonant biotic relations in the Earth system—the primary throughline of this book so far. Organisms in the Anthropocene are doubled in their own environs. Biotic identity ecometonymically references the dissonance of organisms in an age that rapidly reiterated planetary environs. Organisms that integrated with pre-Anthropocene Earth—those evolutionarily manifested by interrelations that largely secured a systemic stability for the planet—are out of place in the very environs they once innately maintained. Potential corrective measures recharacterize anthropogenic impact to a more controlled output of the anthroposphere. The notion that a planet is not for its inhabitants and our responding ecological control look very much like our current and forthcoming extraplanetary lives. Elsewhere in the universal ecosystem, identity is primarily constructed along biotic connections. Maintaining resources such as food and oxygen accentuate biotic identity as pre- or superseding ontological constructions of self— an ecoethic of biotic integration emerges for reinsertion with sociopolitical humanity. Whatever our ecoethic, it initiates the late-stage Anthropocene and guides the planetary-scale efforts to stabilize the biosphere—that reversal of natural selection in which planetary ecology is selected by human physiology. The Anthropocene is very much characterized by biotic identity evidencing planetary ecology.

Biotic identity ecometonymically describes the abiotic features of an organism's environment. While organic perpetuation is the rhetorical catalyst for action in the Anthropocene, the epoch has always pivoted on abiotic and inorganic factors like average global temperature, carbon output, and sea level. The intersection of planetary-scale ecological control and extraplanetary permanence makes clear our species' interrelation with the abiotic features of the universal ecosystem. Maintaining biotic selves in the universal ecosystem exacerbates the environmental dissonance experienced in the Anthropocene. When this discord is resolved, when no environment refuses us, when we integrate with the universal ecosystem to maintain our biotic selves, the Astropocene begins. In the universal ecosystem, biotic identity ecometonymically traces to abiotic factors that feel even less organic or "of nature" than abiotic environmental contours such as average planetary distribution of ice and liquid water or tectonic geomorphology. Unlike the relative ease of connecting terraforming with the Anthropocene's ecological character—the action creates an environment that innately provides the biotic needs of an organism—entrance into the universal ecosystem finds the environmental sciences engaging abiotic factors best described by physics.

This final chapter interrogates biotic identity formed of interrelations in the universal ecosystem that exemplify how we read upon acceptance of the Astropocene's abiotic landscape. This analytical process evidences what it means to be human in the Astropocene so that ecocriticism can apply this foresight to contour the ecoethic that initiates the late-stage Anthropocene. Continuing with *Seveneves* and *The Martian*—nature narratives for the reciprocating ecocritical method—the first close-reading examines biotic interrelations in vacuous environs and notes how humans in *Seveneves* interact with the limited biotic connections in their permanent space stations. Planetary conceptions of environment are challenged by extraplanetary existence, although human ability to describe the universal ecosystem is already well established. Following Watney's lead, entropy in the second law of thermodynamics traces a conceptual pathway that describes biotic human interrelations with all planetary and vacuous areas of the abiotic universal ecosystem. The findings of physics and astrophysics already detail the universal ecosystem, and ecocriticism must utilize such scientific frameworks. Invoking these now is a proactive and agential response to the rapid relevance of physical descriptions of the universe to biotic identity in the late-stage Anthropocene.

In the universal ecosystem, of which Earth is but a part, biotic identity ecometonymically traces to abiotic nature best described by physics. Portending ecocriticism's invocation of theoretical physics for the Astropocene, the clearest human impact line in the geological record seems to be the plutonium fallout from the several nuclear detonations beginning in 1945 and extending into the early 1960s. Loudly declaring the Anthropocene, plutonium 239 is almost exclusively human produced, appearing naturally only in rare circumstances. Regional levels may be "augmented by accidental discharges from power stations, reprocessing plants, and satellite burn-up on atmospheric reentry" (Waters et al. 2015, 51), but the dispersal of plutonium 239 across the planet is exclusively the result of nuclear detonations that "can provide a practical radiogenic signature for the beginning of the Anthropocene" (Waters et al. 2015, 55). While anthropogenic impact is hardly limited to just one action, the ability to point to a period of less than twenty years where a single species drew a fine and relatively uniform line across the planet's surface is disquieting. Still concerned with the character of the epoch rather than its start date, this atomic lens quickly relates biotic identity to the universal ecosystem's abiotic features—an ecometonymic description of human biotic identity in the Anthropocene and Astropocene.

The nuclear detonations are but one expression of a larger conceptualization of abiotic factors that are far more descriptive of the universal ecosystem than any single planetary ecology: "By splitting uranium and plutonium atoms, scientists had made a weapon by using the very same principles that made the sun shine: $E=mc^2$" (Seife 2008, 3). In addition to reinforcing the pathways between anthropogenic impact and the physical, atomic nature of stars, Einstein's special relativity—the theory that conjures that most famous equation describing the equivalence of energy (E) and mass (m)—begins to identify the abiotic contours of the universe. Most notably, the special theory of relativity identifies the maximum speed at which anything can travel, the speed of light. Special relativity, however, does not relate gravity to its equations and thus does not quite assess the universal ecosystem. Not until the theory of general relativity is gravity included, and human conception of the universe is paradigmatically expanded. Einstein's theory realizes the very spatiotemporal structure of the entire universe. The theory finds that "gravity is not a force like other forces, but is a consequence of the fact that space-time is . . . curved, or warped by the distribution of mass and energy in it" (Hawking 1988, 30). General relativity describes how gravity impacts the spatial and temporal dimensions: "Space and time not only affect but also are affected by everything that happens in the universe . . . and it became useless to talk about space and time outside the limits of the universe" (Hawking 1988, 34). General relativity describes interactions with and in spatiotemporal reality. By general relativity, the abiotic feature of gravity ecometonymically connects biotic identity with the very spacetime of the total universe.

The universal ecosystem's natural environs are all governed by the physics that describe the nature of spacetime and the predictable interactions of particulate matter within. Such theories detail biotic interaction with the spatiotemporal nature of the universal ecosystem. Ecocriticism in the universal ecosystem must extend its scientific literacy to include physics, both astro- and theoretical. This chapter invokes thermodynamic descriptions of entropy and general relativity in ecocritical literary theory. In doing so, ecocriticism can, right now, in the Anthropocene, read as though in the Astropocene. This sort of interrogation shapes the late-stage Anthropocene and the Astropocene and the ecoethic by which the epochs unfold—and perhaps even glimpses what is beyond them.

VACUOUS ENVIRONS

Thus far, discussion of survival in the universal ecosystem has been planetary, giving it an organic tangibility and materiality that is characteristically opposite the void of space. Speaking on the dangers of the vacuum, Venkat Kapoor says of the Ares crew in *The Martian*: "Nobody thinks about it, but statistically they're in more danger than Watney right now. He's on a planet. They're in space" (Weir 2014, 127). Planetary ecosystems seem to have a higher tolerance for error, creating a mediating environment that reinforces Watney's survival. Such astropic interrelation destabilizes the internal/external binary, approaching a direct biotic relation with the environment beyond Earth and the simulations that allow entrance into the universal ecosystem. A planet like ours, by its gravity, orbit, moons, and eons of universal interactions, is quite stable relative to human timescales. Watney has a Hab designed to respond to baseline Martian environs, which in turn provides resources for him to work with. He uses Martian dirt and rocks, temperatures, solar energy, and lunar cycles to sustain his earthly body. Even the lack of atmosphere and gravity help Watney reunite with his crew. Watney's survival is planetary, even while not so earthly.

Space is so totally uninhabitable that it seems not to register with the organic roots of humanity. Where Watney can draw on external resources, his crewmates returning home aboard the Ares have next to no resources beyond the hull of their ship. What they have is what they bring with them. Their biotic characteristics ultimately limit the ship, whose travel speed is "defined not by the ship's power, but by the delicate human bodies inside" (Weir 2014, 142). Even the benefit of the frictionless vacuum is subsumed by the human ecometonymy formed by Earth's baseline gravitational forces. At their most interrelated with the universal ecosystem, the Ares crew uses Earth's gravity to slingshot back to Mars to rescue Watney. This is a classic maneuver closely related to the Astropocene, though only a proto-astropic act as it is not using gravity in the biotic formation of humans but rather as a sort of fuel, inexhaustible but still anthropocentric. Such maneuvers signal an entrance and relation to the universal ecosystem, incorporating gravity in terran actions. The maneuver saves Watney from dying on Mars but is sociopolitical rather than biotic.

Vacuous environs are far more difficult to integrate with than planetary environs. Watney has additional resources that are simply nonexistent for the survivors in *Seveneves*. Looking at the invocations of nuclear reactions as critical to movement within the universal ecosystem, the Ares crew just buries their RTG on Mars once they arrive. Watney digs it up and uses it to balance the climate of the Rover with that of Mars while he travels across the surface. Afterward, he just buries it again (Weir 2014, 74). Watney's survival on Mars is very much a planetary act with ecocritical significance. It models biotic relation to an environment, which allows greater margin of error by virtue of planetary location. Terrans in *Seveneves* must live near their reactors, repairing and salvaging spacecraft that malfunction—reclaiming a ship with a cracked reactor and radioactive particles invisibly flitting around the ship becomes an ecogothic tale for the universal ecosystem, a terrifying and visceral amalgam of nuclear fallout, the contagion narrative, the inhospitable wild, and the hidden monstrosity on the space station burrowing out from human bowels (Stephenson 2015, 371–389). Watney has fewer options for surviving an RTG leak than terrans in *Seveneves*, but his overall biotic self is maintained by the Martian environs. Even an inhospitable planet provides enough abiotic substance for Watney to maintain his biotic self, acting as a mediator between the hab and vacuous environments. Unlike *The Martian's* focus on astropic intersections of planetary ecologies, living in vacuous environs accentuates the interactions described by physics as human biotic relations are restricted to the minimum earthly environs maintained by a ship's hull.

Echoing Venkat Kapoor's statement, the critical disasters of the space-faring humans in *Seveneves* almost always kill someone. Watney has a planetary layer between himself and space. With it, he approximates earthly soil and grows a potato crop to again model increased opportunity in his biotic relation with a planet. Unlike the planetary Watney, survivors in *Seveneves* are biotically restricted with only vacuum outside their earthly environs. Objects are dangerously moving through the universal ecosystem, and the potential for harmful collisions is dramatically increased as the survivors remain in Earth orbit near the colliding lunar debris. Watney's biotic relations are characterized by embedding a static habitat in the landscape while the Hard Rain survivors are constantly moving through space. Afterall, they are Spacers while Watney is the Martian. Being in orbit means constant movement and adjustments for drift, circling but rarely benefiting from the planet, before even having to deal with incoming lunar debris.

Biotic in approach, the Spacer's incongruous entrance into the extraplanetary environment riddled with lunar debris selects the species' behavior in the universal ecosystem. Before settling on a large lunar fragment for some planetary benefit, the first iteration of the spacer colony formed the Cloud Ark, a highly mobile swarm of habitats called arklets. The arklets are launched as frequently as possible, maximizing humanity's off-planet population. Once in orbit they approach the ISS and dock with other arklets, creating small communities connected to the whole by a digital network. The Ark Colony's survival strategy is to amass the ark groups in roughly the same location. Since they are orbiting Earth with a moon's worth of rocks that could easily destroy an arklet group, they prioritize monitoring incoming debris. When the trajectory is identified, the arklets disconnect and maneuver out of the way as the rock passes through the colony, coalescing back into the group afterward.

The action is modeled after studies of swarm behavior, the decentralized but cohesive mechanics by which a school of fish moves together until separating so a predator passes harmlessly through their midst or how a flock of birds appears to act as a singular organism (Stephenson 2015, 34–35). In this construction, they constitute a terran cyborg, in which they technologically adopt the organic mechanism of other species, akin to Watney's plant-like use of solar panels to direct his movement, affirming the inclusive terran nomenclature and a type of ecometonymy. As the only organism beyond bacteria and algae to survive the Hard Rain, adopting the action of another species references the biomechanics and biotic relations of species not likely evidenced in a fossil record. The biotic interaction with the universal ecosystem traces back to an algorithm informed by swarm mechanics as observed in several terran species preceding humans, perhaps developed by some of the first swimming prey organisms.

The survivors interacting with their environment as swarming earthly organisms does not mean there are enough resources. The ark groups are largely self-contained, and the extreme circumstances leave each subgroup to fend for itself. Similar to Watney, in extremis entanglement in the vacuous spaces of the universal ecosystem forces terrans into strange biotic relations. Mars allows Watney to maintain nature's indecipherability of that which is internal and external. Earth and Mars astropically entangle to create an Earth system for Watney. The Cloud Ark colony looks far more like classic science fiction than Watney's farm, as they are growing algae designed to provide oxygen, nutrients, and calories (Stephenson 2015, 186). Biotic systems become only as big as the arklet couplings and some are unsuccessful at maintaining their food supplies. Without the ability to draw on external planetary resources, such containment still creates biotic interactions, but nature as a system, always already without the internal/external binary, exacerbates instead of perpetuates terran survival.

In the landscape between planets, with no organic environment to draw on, the Ares crew establishes protocols for cannibalism and nature gains the ominous, darker notes of the void, as the body biotically links to the only other organic matter: the other human bodies aboard the ship. In *The Martian*, Watney's original crew slingshots around the Earth for a gravity assist, returning to Mars for his rescue. While passing by Earth they will dock with a resupply probe containing what they need for another trip to Mars. Johanssen, the Ares computer and systems specialist, has a brief conversation with her dad as they pass by the planet. She assuages his worries by asserting she will live, even if the resupply fails. She explains that the crew designated her their survivor. The smallest of them, she needs the least food and has the necessary skillset to return the ship to Earth. The rest of the crew would immediately imbibe lethal amounts of pills, so they take up no further resources. Johanssen explains that their one-month food supply turns to nine if she rations, adding that the trip is seventeen months. Her father asks how she survives. Johanssen ominously offers, "The supplies wouldn't be the only source of food" (Weir 2014, 253).

Johanssen is ecometonymically describing the Earth system by her dietary needs. The cyborg human in the late-stage Anthropocene resolves the disconnect by simply bringing supplies with them, getting resupplies, and eventually returning to Earth. The failure of the resupply probe severs a necessary connection to Earth's resources, and the Ares goes from a simulated earthly environ that sustains its passengers between planets, needing only to maintain the barrier against the void, to the entire biotic environment. Unlike Watney, Johanssen will not simulate a full Earth cycle by approximating Earth soil. She could halt the decay of her new food supply by shutting down the temperature controls in whichever area she designates as freezer, including the thermal characteristics of her vacuous environs into her own. This refracts Watney's astropic act of storing potatoes on the Martian surface, although far darker and much less palatable. Other than temperature that impedes entropic decay, there is little the environs surrounding the Ares can offer Johanssen. An extreme scenario but anticipated from the start, such cannibalism is exactly the kind of apolitical biotic priority initiated by entrance into the universal ecosystem. The resupply is successful in *The Martian*, but in *Seveneves* there is no planet from which humans can draw resources to suit their biotic needs. They either produce earthly resources within the spacecraft, or they seek sustenance elsewhere in the only ecosystem they have. The survivable environment extends only to the spacecraft one is inhabiting and those with which it is docked.

Nature's cyclical intolerance of the internal/external binary continues as the Cloud Ark ecosystems fail to produce food. The arklets maintain a simple ecosystem centered on algae blooms as food and oxygen producers for the terrans on board. For years the survivors are unable to escalate their agriculture beyond algae. Even the rice they develop for low gravity growth needs some semblance of up and down for its root structure to develop (Stephenson 2015, 287). Agriculture eventually crashes on the arklets and cannibalism begins. The initiating act is termed "soft cannibalism" (Stephenson 2015, 520), as a terran eats his own leg because those limbs are unnecessary in space. The terminology indicates a process of indoctrinating the act into social practice after a reversion to the biotic self. The physiological need of food and the environment's lack of gravity coalesce to describe that an organism without legs is no less efficient than one with, freeing the limbs for consumption.

Unexpectedly and proactively biotic, the consumption of limbs mirrors and resists catabolysis, the body's natural process of utilizing muscle protein after fat reserves are depleted. Unnecessary limbs go from external to internal to maintain the full bodily system, of which there is now less to sustain. There are also plenty of people who are already dead, who have no use for their body. Hunger in the universal ecosystem displaces ontological concerns and recirculates to construct an ontological terran accepting of cannibalism. Doubled in the Anthropocene and again in the universal ecosystem, biotic identity fundamentally challenges ontological constructions of self that are luxuries of terrans on our biotic home. This soft cannibalism demonstrates the biotic existence that pre- and supersedes ontological constructions of self. In one sense, the non-ontological identity is short-lived, quickly softened for palatability. The widespread, rapid acceptance of cannibalism in the Ark Colony alternatively suggests that the biotic relation to the only organisms in the environment signals that biotic circumstance is the primary factor.

In the SF feedback loop, the cannibalism episode of *Seveneves* terrifyingly resembles food inequity among global populations. The Cloud Ark resorts to cannibalism while those on the space station only consider it, instead eject-ing their freeze-dried dead into space (Stephenson 2015, 20). Hopefully, we in the Anthropocene can manage without starvation turning to cannibalism. I am far less optimistic about climate change, population increase, and resource inequity resulting in anything other than violent conflict (Mach et al. 2019). Aïda, the Eve who pronounced the curse extending from and as racism, leads the Cloud Ark on an attack against the privileged space station. Her group killed and ate the man who initiated soft cannibalism so they could be well fed for their assault (Stephenson 2015, 535). With reciprocating ecocritical action, the timing of the Arker's assault indicates a deep relation to the uni-versal ecosystem.

The survivors in *Seveneves* are entrenched in the near-Earth vacuous environs of the universal ecosystem. The environment is most immediately defined by the orbital movement around Earth with the lunar debris that inspired their adoption of swarm mechanics. Since the beginning of the novel, the space station is coupled with an asteroid, and the survivors have been using it as a shield against radiation and incoming objects. Similarly, a major venture of the survivors was to send a crew from their geocentric orbit to a heliocentric orbit to intercept and collect an icy comet core that can be turned into hydrogen peroxide for propellant or split it into its component atoms for use in the larger engines (Stephenson 2015, 126–127). They cover their crafts in ice to shield against solar radiation while still letting light in for photosyn-thesis (Stephenson 2015, 503). The ultimate plan is to settle the species into a canyon on a large metallic fragment of the moon, safe from the threatening contours of the universal ecosystem's vacuous environs.

During their approach to the lunar fragment, there is a coronal mass ejec-tion. CMEs are a release of charged particles from the sun, whose negative effects are mitigated by the Earth's magnetic field (Stephenson 2015, 507). They are expected by the survivors and are constantly monitored. The sur-vival procedure for a CME is to enter a sleeping bag-like container whose walls are filled with water to absorb the high-energy protons emitted and take a drug to protect DNA against the radiation. The procedure is fairly simple, though it requires resources that are not necessarily abundant. During the approach to the lunar fragment, the Arkers will dock with the ISS to settle together. During the CME though, the Arkers attack, thinking that they would have an easy takeover with everyone in their shelters (Stephenson 2015, 538). The assault was unsuccessful because a data dump from the arklets occurred and a person who had been knocked free of the station while on a spacewalk was checking the communications data and alerted the ISS over comms that the Arkers were preparing to attack (Stephenson 2015, 533–536). He floated, waiting to die in vacuous environs, preserving his friends as a cyborg terran in an inhospitable landscape. But for this, the Arkers would have easily succeeded by utilizing the CME to their advantage.

Entrenched in the vacuous environs, the survivors do their best to biotically relate to the universal ecosystem. They gather resources, find shelter, and even allow their actions to flow with the contours of their unusual environment. These biotic interrelations are yet another doubling, that state of having a body suited to another environment. Humans, evolved to singularly suit our biotic home, so the mediating environments produced by spacesuits, habitats, and ships are cyborg constructions enabling our entrance into universal environs incompatible with our terran anatomy. Elsewhere in the universe, the biotic terran identity ecometonymically describes Earth by the contrasting absence of necessary and innate features. Biotic interrelation is never halted though. An organism always exists in relation to its environment. An environment such as the vacuum of space may be completely uninhabitable by the entire terran biosphere (although that may not be true as tardigrades and other extremophiles may not die), but a death in a vacuous environment is still natural selection.

Natural selection occurs everywhere in the universe, but it seems a poor descriptor for terrans experiencing the sheer impermissibility of the void or any other planet that is even marginally different from Earth. We are unable to survive most places in the universal ecosystem, yet we will live in them. The Astropocene is characterized by unprecedented biotic relations to novel ecosystems. Such a biotic identity for our species inculcates a paradigmatic shift in concepts of nature. All environments and organisms, all biotic and abiotic characterizations, all iterations of nature ecometonymically describe spacetime. The physical laws describing the particulate and material processes of the universe are the abiotic contours of biotic identity in the Astropocene.

ENTROPY

The search for underlying laws of the universe is the pursuit of physics and its various interdisciplinary manifestations. The most dramatic expression of the pursuit is the search for the "theory of everything" that would resolve the apparent incompatibility of general relativity and quantum mechanics.

Particle physics has already unified the equations describing the irreducible fields that constitute the electronuclear interactions of particles. Gravity is the only known remaining field that interacts with matter in the universe. Unifying our gravitational and electronuclear equations should make calculable, that is predictable, all physical interactions. Quantum mechanical uncertainty limits the utility of such an equation to predict highly variable futures like lottery numbers or what word I will say next. The theory is for understanding things that pattern the universal environment. A patterned and predictable equation reconciling quantum mechanics and gravity is nonetheless implied by everything we know about the orderly structure of the universe and will reveal dramatic information about the initiation and end of the universal ecosystem. You should experience a sense of imminence surrounding the equation. We have several theoretical approaches to a theory of everything, and they may each even be correct in their way, respectively describing the universe from a certain point of view (Hawking and Mlodinow 2010).

For now, gravity remains disconnected from the other fundamental fields governing interaction in the universe while, in and of itself, the effect of gravity on the dimensional fabric of reality established in general relativity is well understood. Quantum mechanics signals an incompleteness of general relativity as a model of physics but this should not be taken as the unequivocal failure of general relativity. Einstein's exhilarating descriptions of spacetime and gravity offer astounding visions of the universe, and the claims in general relativity are consistently verified by experimental physics. Einstein's equations detail that understanding gravity is to understand the inseparability of space and time, the four dimensions within which the existence of anything is formed. All things exist within the spatiotemporal fabric of reality. The physical laws describing how relations in spacetime unfold describe biotic relations in the universal ecosystem.

All matter and energy reside in these four dimensions. What we know about the origin of the universe indicates it began with its entirety compressed into a single point from which expands the cosmic structure we see in our telescopes and equations. The universal ecosystem settles as black holes, suns, and planets coalesce and, by their gravity, curve spacetime. The curvature of the spatiotemporal fabric, expressed as gravity, draws less massive objects into collisions and orbits with their neighbors. Some objects are moving too fast and will escape their immediate gravitational environs and find their way elsewhere, eventually drawn into some system. Black holes are orbited by galaxies of stars who are in turn surrounded by planets and their moons, asteroids, comets, and whatever other marvels move within the universe.

Stephenson describes that our solar system was previously much more disordered. Once, our "solar system had been a much more chaotic place, with a wider range of orbits, but the processes mentioned had gradually swept it clean and, by a kind of natural selection, produced a system in which nearly everything was moving in an almost circular orbit" (Stephenson 2015, 337). Objects had greater variance in orbital trajectory, but collisions and ejections ordered everything into the wide elliptical orbits of the current system. As the sun's gravity draws objects to it, their straight trajectories become solar orbits. Some objects are jettisoned as they fail to maneuver the tight turn of their ellipse while others are rent by tidal forces of gravity or destroyed by thermal effects as they move too close to the sun. Asteroids may find their way into geocentric Earth orbits but are largely snuffed out as they impact the moon. Eventually, this solar system achieves gravitational equilibrium and the surface of our planet turns blue.

On this or any orb, all organisms evolve within the contours of gravity and myriad other abiotic factors. This is not to say that natural selection is occurring to shape the solar system. Stephenson offers that the solar system forms by a "kind of natural selection" (Stephenson 2015, 337). Specifically, a form that does not require biotic relations; the universal ecosystem does not necessitate some vacuum-breathing leviathan exist. By this, we benefit from the distinction between *develop* and *evolve*. The solar system *develops* in response to gravitational forces, while terrans *evolve* in response to gravitational forces. *Seveneves* shows that plant systems evolved in line with Earth gravity (Stephenson 2015, 287), and *The Martian* even explains how, in Martian gravity, skipping moves a human across distances more effectively than running (Weir 2014, 167). We can move quickly on Mars, but we are running animals biomechanically evolved in Earth gravity. The universe becomes ecological only by the interactivity of organisms and environments. There is enough exchange between Earth and the universe to characterize a universal ecosystem without terrans leaving the planet—just ask the dinosaurs about otherworldly objects incorporated in the environment. The Astropocene identifies the universal ecosystem in which terran biotic relations cannot be described by the geology of one planetary object, whether it be a missing comet, a terran body on a planet from which it did not originate, or a once red planet is now blue. The thematic parallels between development of the universe and evolution of a planetary system are admittedly enticing, especially as we consider such thermodynamic equilibrium from our place in the climatic imbalance of the Anthropocene.

While a sort of natural selection without biotic relations must suffice, lifeless or not, the underlying concepts of particulate interactions within the ecological or abiotic universe are constant. Considering the universal ecosystem is to consider the underlying spatiotemporal relations, especially as those physical descriptions are constant while organic systems are not guaranteed. The equations of physics, by their very definition for they are incorrect otherwise, mathematically describe the laws governing the consistent interactions of all things throughout the universe. According to Einstein, the laws of thermodynamics are the physical descriptions least likely to need revision (Einstein 1969, 33). As these laws describe equilibrium and the interconnectivity of macro and microsystems, they are a fine place to begin the expansion of ecocriticism's scientific literacy for describing terran relations to the universal ecosystem.

Thermodynamics describes the transfer of energy, especially heat, among connected systems. Those picking up this book will likely find the first law familiar as it is an expression of the law of conservation of energy: energy cannot be created or destroyed, only changed in form or location. Thermodynamics' first law tells us that a sufficiently isolated system maintains an average equilibrium, meaning that as the system unfolds the amount of energy is transferred and transformed but never exits the system. As far as we can tell, the universe does not connect to anything else and is thus a macrosystem in equilibrium, altering only the state of energy, matter, and mass, never the total quantity (Berry 2019, 14–15)—again nature disrupts the binary. The universal ecosystem does not gain or lose any fundamental components, although the microsystems within may lose or gain energy. Consider the prior example of an object ejected from a solar system. The kinetic energy of the object exits heliocentric orbit, leaving that microsystem of the universal ecosystem and is transferred to another system, colliding with something in it or finding a sustainable orbit.

Deceptively succinct, the second law states that a closed system increases in entropy. Entropy is disorder, most usually conveyed as heat, the vibration of atoms. The fusion reactions of the sun emit radiation that, roughly eight minutes later, transfers energy to the molecules in the air and your skin, increasing heat energy or entropy. Or, imagine again that object exiting a solar system as its velocity was too great for it to make the turn of its elliptical orbit—say, an asteroid, dense and metallic. This asteroid exits its solar system and enters ours. Our solar system has increased in disorder, in entropy. The original system loses energy and matter while we gather more. The universe remains in equilibrium and our solar system, though measurably more disordered, is relatively unaffected.

Recall now the Agent in *Seveneves* and imagine that the asteroid was sufficiently massive and traveling with extraordinary velocity toward Earth's moon. Imagine the impact transferring the Agent's energy into the moon and the moon exploding, shattering into seven separate pieces. Some shrapnel will exit geocentric orbit to impact other cosmic bodies, in this and perhaps further systems. Perhaps the Agent does this too. These lunar pieces collide and grind, creating a beautiful and disordered set of rings around the planet. This is a far more noticeable entropic increase. The universe remains a closed system, marking no overall change but the individual microsystems of the moon, of the Earth, and of whatever is eventually impacted by those pieces that set off about the universal ecosystem distinctly increase in entropy. As time progresses, a closed system always progresses toward maximum entropy (Berry 2019, 87). At maximum entropy, the closed system achieves thermodynamic equilibrium and never reverts to a state of lower entropy without work being done (Berry 2019, 43).

Entropy is a description of the universal ecosystem. The first law of thermodynamics describes the constant energy of a closed system, such as the universe. The second law describes how thermodynamic systems always approach their maximum entropy. There is a zeroth law of thermodynamics, preceding the others conceptually although not historically, and establishes the transitive property among thermodynamic systems. This law is incredibly important for relating systems, especially by temperature as a measure of entropy and equilibrium, although it is far less fun than the others. The third law of thermodynamics describes minimum entropy at zero kelvins, better known as absolute zero.[1] In ecocriticism, entropy acts as a feature of the universal ecosystem to which terrans biotically relate.

Watney's use of the RTG is a study in biotically relating to entropy, and he both finds it "nice to see thermodynamics being well-behaved" (Weir 2014, 234) and specifically curses entropy. The decay of plutonium in the RTG—which Watney notes is a "natural process happening at the atomic level" (Weir 2014, 73)—radiates heat that is transferred to the air molecules in the rover, increasing the entropy of the external system to simulate the earthly climate. The heat energy produced is too much for Watney. He detaches a piece of the insulation to astropically invoke the Martian temperature to balance the extreme heat generated. While no insulation is perfect, the heat increases the entropic vibrations of the rover's molecules which in turn transfers heat to the Martian atmosphere. He still needs to reduce the temperature of the rover system by facilitating the transfer of energy to the Martian system.

Without the RTG, the rover's heater would drain too much power for him to travel as quickly as he needs. Before he dug up the RTG, he tried to use his own body heat to create a thermodynamic equilibrium between his body and the rover. The rover has top-notch insulation, best on Earth, but Watney's plan quickly gets outstripped by the rover's thermodynamic connection to the Martian atmosphere. Watney curses, "All my brilliant plans foiled by thermodynamics. Damn you, Entropy" (Weir 2014, 72). Watney frequently asserts that Mars is trying to kill him but here, he traces his biotic relation to physical, natural law. Watney is biotically entrenched in the universal ecosystem which does not appear biotically or even abiotically uniform. The universal ecosystem is only consistent when one relates to the fundamental physical laws. Watney is no longer concerned with the planetary system but is biotically interacting with entropy, with the universal ecosystem.

In order to predict and shape the next geological epoch, ecocriticism must include physics in its vision of nature. Again borrowing Haraway's phrase, ecocriticism needs to become "mathematical and fleshy" (2016, 120) for that is what we will be in the universal ecosystem. Biotic identity in the Astropocene is derived from a relation to unprecedented physical locations exerting their novel abiotic contours on terran bodies. The humans of *Seveneves* become a cyborg swarm by the mathematics describing fish and birds naturally swarming. Watney's body experiences entropy on Mars as it would everywhere in the universal ecosystem. Extending scientific literacy to include physics, ecocriticism is a profound and robust technology for describing humans in the Astropocene.

ECOCRITICISM OF THE UNIVERSAL ECOSYSTEM

Locating Watney's biotic interrelation with the physical dimensions of nature, ecocriticism realizes the ability to utilize and describe the expansion of nature to the total universe. Engaging the universal ecosystem requires precise understandings of the fundamental interactions of the universe, such that any terran entrance to an extraplanetary environment is also a biotic relation to the spatiotemporal fabric of reality governing the production of abiotic and biotic features. Treating the environment with the grand pervasiveness of the universe illustrates the uniform characteristics by which all things manifest and is not meant to displace local biotic relations. The uniformity does not replace anything, enhancing the scope and potency of ecocriticism at the end of the Anthropocene.

Opposite the sprawling scope of the earlier chapters, I turn to *The Time Machine* by H. G. Wells to perform three readings of the novel as a comparative case study of ecocritical perspectives that illustrates ecocriticism's incorporation of the universal ecosystem to critically frame the Astropocene. I read *The Time Machine* three ways, beginning by connecting the social allegory of the traditional reading with Nixon's slow violence. The second reading employs a Darwinian frame where second-wave ecocriticism focuses on the biotic relations in natural selection to find the doubling characteristic of the Anthropocene. Finally, ecocriticism for the universal ecosystem reads, by still following entropy, the evolution of humans on Earth and the novel's solar death as the same entropic decay, all entwined in one of the first articulations of time as the fourth dimension. These readings are non-mutually exclusive, instead stacking and maintaining *The Time Machine*, establishing ecocriticism as a commanding perspective for the environmental humanities, parsing texts to inform biotic relations in the next geological age.

Social Allegory

The foremost reading of *The Time Machine* focuses on an evolutionary divergence of the human species into Eloi and Morlocks as a social allegory for the capitalist-laborer split of industrial London. The SF classic follows the Time Traveller as he uses his time machine to explore the future. In the SF feedback loop, the novel details the evolutionary impact of factory labor and its capitalist benefactors as the Time Traveller encounters their evolutionary counterparts, the Morlocks and the Eloi. The Morlocks appear brutish albino apes, nocturnal for they dwell underground in deep tunnels. Extending from the gritty laborers in factories and mines, they are physically strong and technologically capable. They maintain if not engineer the mechanisms by which oxygen is brought into their cavernous domain (Wells 2001, 115). The Eloi descend from the capitalists who without work enjoy the picturesque countryside, landscaped to delight. Exponentially dainty and frail, the Eloi caricature the capitalists. The Eloi frolic and copulate, carelessly plucking flowers and eating of nature's abundant resources. They even retain no knowledge of pockets and surmise that the Time Traveller's must be for carrying flowers (Wells 2001, 121). At night, they sleep huddled together in dystopically dilapidated buildings afraid of the Morlocks roaming the hillside and forest. By this allegory, Wells posits a devolution of the species as capitalism systemically disrupts the intellectuality the Time Traveller believed would extend from Victorian London. Instead, future speciation proceeds along class lines by slow violence.

The Time Machine reduces the felt experience of human individuality and variance into two archetypes—a powerful refraction of capitalism, even now. The Time Traveller, not the novel, then filters out the nuance of the Eloi and Morlocks by being more intellectual, which is not to say that he has a more robust culture or that they are out of balance with the ecosystem. Anthropocentrically, he never realizes that nature does not inherently select for intelligence. The Time Traveller passingly wonders about the production of Eloi clothing when they seem to have no facility or practice of production or repair, noting their sandals are "fairly complex specimens of metalwork" (Wells 2001, 102). He never realizes the full scope of the Morlock's cultivating relation to the Eloi population. The anthropocentric vision of simplicity reduces their interactions to predator-prey dynamics that mask the complex stewardship that is akin to the wolf and antelope otherworld opening McCarthy's *The Crossing*. The Morlocks have converted their material production for capitalism into a sustaining practice that perpetuates biotic relations by clothing the Eloi.

With *The Time Machine*, Wells provides Victorian imagination with novel connectivity between the destabilizing extensions of evolution. Evolution by natural selection

> challenged the belief that humans are at the center of the world, positing instead that humans and animals exist on a continuum. According to Darwinian tenets, both descended from common ancestors and therefore share many common tenets. The recognition that species become extinct, and that more organisms are born than can survive led to pessimistic views on the inevitability of a painful struggle for existence and to optimistic ones applauding progress and improvement. (Anger 2014, 55)

The Time Machine disrupts humanity along the entirety of this continuum. Darwinian thought unravels differentiation, especially by divine creation, between terrans. Natural selection implies that humans arrived in the present by the same means as everything. Speciation occurs by the utility of random mutations in the environment. Wells assaults the present with an evolutionary extrapolation that opposes notions of improvement, threatening that society is on a trajectory of degradation.

The environmental humanities does not even slow down for the Wellsian allegory, currently identifying the effects of capital and industrialization on a global scale, scientifically in the environmental impact of humanity, particularly the global north, and ecocritically in the narratives of those regionally and globally marginalized. Rob Nixon details the process of slow violence, "a violence that is neither spectacular nor instantaneous, but rather incremental and accretive, its calamitous repercussions playing out across a range of temporal scales" (Nixon 2011, 2). Structuring and harming the environment always favors privilege, especially as granted by capital, by unfolding slowly against a community without the power to resist. Consider, for example, the act of burying toxic waste in an impoverished country or the placement of a power plant near those urban residents with the lowest income.[2] The destructive impact of environmental violence, that is violence against and through the environment, can take generations to manifest in an individual's biotic identity.

Such environmental catastrophes "that overspill clear boundaries in time and space are marked above all by displacements—temporal, geographical, rhetorical, and technological displacements that simplify violence and underestimate, in advance and in retrospect, the human and environmental costs" (Nixon 2011, 7). *The Time Machine* sees the entire working class at some point moved underground, living and working in the same subterranean space. They are displaced from the very environment needed to survive by the efficiency of capital. Air is pumped down to them, cruelly forming cyborgs who only reclaim an unmediated relation to Earth's atmosphere on an evolutionary scale of time. The Morlocks become physically intolerant of light and fire and are biotically suited to nocturnal surface activity.

Slow Violence and the Environmentalism of the Poor recognizes that environmental violence is "a contest not only over space, or bodies, or labor, or resources, but also over time" (Nixon 2011, 8). If one thing can be said about the primal and visceral future of the Eloi and the Morlocks, it is that the ultimate irony is the Morlocks eat the Eloi (Wells 2001, 124). I am uncertain if the speciation is divisive enough for the image to discontinue the cannibalistic motif of this chapter, but it certainly dismantles consumptive capitalism with natural process. The ironic hope of *The Time Machine* is its moral, a reversal where the worker finally consumes the capitalist while reciprocally maintaining their population by material production. The social allegory of *The Time Machine* functions by the process and scale of slow violence along delineations of capital and labor. Ecocriticism innately and adroitly derives the allegorical intent of *The Time Machine*.

Darwinian Lens

The second reading of *The Time Machine* employs ecocriticism's extrication of nature from metaphorical representation to favor a Darwinian reading of *The Time Machine*. By the time Wells publishes *The Time Machine*, Darwinian natural selection is well established.[3] Wells is familiar with the processes by which nature selects the physical adaptations of all species. During the late nineteenth century, biological degeneration was believed to be a viable outcome of natural selection, opposite that of sentience—adding to the allegory that Morlocks while intellectually deteriorated remain far more intelligent than the Eloi. This idea travels directly to Wells from Thomas Huxley and E. Ray Lankester to become his notion of zoological retrogression (Ruddick in Wells 2001, 162). The idea that the whole of humanity could fail to retain elevated sentience and end up equally degraded as the Eloi and Morlocks was a harsh and relevant criticism of Victorian London, although ultimately flawed.

The concept of zoological retrogression is distinctly anthropocentric, extending that humans are superior organisms by virtue of intellect. Admittedly, one would be hard-pressed to say that intelligence is not widely useful in terms of securing food and escaping predators. However, there are an abundance of organisms without brains who are doing just fine and have been since well before humans. Surprisingly, eschewing the anthropocentric lens of zoological retrogression in favor of pure and chaotic natural selection does not make the divergent speciation less likely.[4] Wells creates an ecosystem replete with organisms manifesting random mutations that may not matter for generations, maybe ever, and a landscape that can shift and cease to benefit an organism once favored. Wells correctly images natural selection even while he anthropocentrically misinterprets its trajectory in order to threaten capitalists.

The allegorical application of a concept is the hallmark of the humanities and an inherent output of the SF feedback loop. I suppose it is not even problematic that human art would manifest anthropocentrism. Second-wave ecocriticism's Darwinian lens unravels the anthropocentric notion to arrive at an equally important perspective that does not invalidate the allegorical reading. Evolution is still a process occurring by biotic relation, the interaction between an organism and environment. Morlock physiology develops to suit the requirements of their environment and the Eloi develop to suit theirs. The Eloi seem to develop a certain ineptitude, even failing to maintain the strength to swim (Wells 2001, 102). The lack of additional predators beyond the Morlocks strikes an odd picture, although the Eloi extend from the capitalists who landscaped the country, routing the predators. Former industrial space is where the two groups currently reside as the Morlocks are a far more static group.

Stripped of the anthropocentrism that imagines the human centrality eventually manifesting as the terraforming reversal of natural selection, culture is tinted with the notion that humans have extricated themselves from natural selection, from nature, by virtue of the civilized boundary. We know that nature has no internal/external boundary. Especially after equating it with spacetime, we know that it has always already made such notions irrelevant. *The Time Machine* functions by virtue of cityscapes as environments naturally selecting—which we know them to be. We have unprecedented control over the contours of our environments, but we never exit nature and natural selection. *The Time Machine* is a display of urban ecologies, those ecosystems circulating within the cityscape, and begins to look like the doubling inherent in the Anthropocene.

The Eloi and Morlocks are physiological evolutions produced by the industrial urban ecologies. While industrial England may not be a likely candidate for the start of the Anthropocene, its prototypical impact is entwined in the character of the epoch. The Eloi and the Morlocks act as doubles of humanity here but they have found balance. We however are imbalanced with our environs and extinction seems more likely than becoming Eloi or Morlocks. There is some change that needs to occur for survival, and it seems to be a technological and wild lifestyle that replaces the cityscape and its capital system. Changing the form of nature does not undo natural selection. We are still being selected for and against. Even missing the mark on evolution, *The Time Machine* provides an ecocritical reflection of industrial civilization that travels forward into the Anthropocene and currently manifests in examinations of urban ecologies. The Eloi and Morlocks are not doubled, the Time Traveller is: "I felt hopelessly cut off from my own kind—a strange animal in an unknown world" (Wells 2001, 96).

In the Universal Ecosystem

Following entropy as a marker of biotic interrelation with the universal ecosystem deepens the social allegory and Darwinian readings for use in the Astropocene by connecting the two with Wells's heat-death imagery of the sun's entropic decay. Tracing entropy through the prior two readings draws in the apocalyptic scene of solar decay witnessed by the Time Traveller as he moves far into futurity. Reading these scenes with entropy reveals the spatiotemporal environs of *The Time Machine* and relates the text to modern physical descriptions of biotic relations with spatiotemporal environs. The study of thermodynamics is invigorated by the need of seventeenth-century tin mines in southern England, framed as a search to mathematically determine the minimum fuel to run water pumps (Berry 2019, 49)—resonating with the pumps the Morlocks use to get air into their tunnels. The second law of thermodynamics is well established by Wells's time. Lord Kelvin, who dies the year *The Time Machine* is published, restates the second law, following Rudolf Clausius's reiteration after its establishment by Nicolas Leonard Sadi Carnot (Berry 2019).

In the social allegory, Victorian anxiety about the futuristic visions of their scientific findings entwine natural selection and entropy. Evolution by natural selection disrupts anthropocentric notions of originating divinity, control over the landscape, and the future of civilization. Wells continues to harry the Victorian imagination by bringing entropy to bear on the sun and its orbiting planets. In 1865, Rudolph Clausius interprets the second law of thermodynamics as entropy to denote the constant increase of a closed system toward entropy (Berry 2019, 87). The disorder always increases. In the wake of entropy, "Victorian physicists offered contending predictions of the years remaining before the solar system would have exhausted its energy" producing "popular anxieties about the heat-death of the sun" (Anger 2014, 62). These pronouncements further destabilize imagined human futures of environmental control encultured via the landscaped picturesque countryside (Williams 1973, 128). Wells's invocation of Victorian era science to prescribe the evolutionary timescale and the solar death ecorealistically frames with the more speculative, if not fantastical, Eloi-Morlock epicenter of *The Time Machine*. Without second-wave ecocriticism though, this dual invocation of natural selection and thermodynamics does not coalesce into a singular conceptualization of the environment.

Tracing entropy in the Darwinian reading reveals the biotic impact of thermodynamics in evolution unfolding in the universal ecosystem. Escalating the allegory, the Darwinian reading details the biotic connection between humans and urban environs, further destabilizing anthropocentric notions of extrication from natural selection. Similarly, ecocritically prioritizing the image of nature as real nature (Parini 1995), reading entropy in the Eloi-Morlock extrapolation, connects earthly organisms to the universal ecosystem by an evolutionary entropy. The Eloi-Morlock split might be interpreted as a reduction in complexity, which the moral of devolution superficially encourages, resulting in an anthropocentric simplification that appears to run counter to the entropic increase of disorder. The Eloi and Morlocks appear quite orderly. If we seek only to preserve the existence of the second law of thermodynamics in *The Time Machine*, we have only to recognize that organisms and planets are not closed systems and thus order appears while disorder appears elsewhere, perhaps later in the disorder of the disrupted planetary orbits. While nodding to the thematic parallels of human decline and heat death is sufficiently cohesive as narrative, the reading is a limited application of extending ecocriticism's scientific literacy to include physics in recognition of the spatiotemporal universal ecosystem.

Reading human evolution and its planetary disarray as increases in disorder aligns the biotic and planetary environments as descriptions of universal nature's spatiotemporal dimensions, the recognition of which is prolifically bound up with *The Time Machine*. In some ways, the species' divergence increases entropy by making two from one. Humanity is now spread across more biotic interactions and is engaging more biotic processes than otherwise. The intermediary civilizations between Time Traveller and the Eloi and Morlocks are evidenced by their decaying architecture. The deterioration of the ordered buildings is an entropic increase in disorder synchronous with the speciating split—both processes of nature. As the world increases in entropy wild nature seems to increase as well, both external and internal to humans. The order asserted over the landscape did nothing to halt the mutative disorder at the heart of evolutionary process that dissolved civilization from within and allows wilderness to consume it from without. Relegated to the subterranean and countryside environs, epigenetic expression would moderate biotic relations to the environment while naturally selected random mutations create the genotypes that become the Eloi and Morlocks. An expression of entropy, a mutation occurs by the failure to perfectly copy DNA.

During this time, population dramatically declines, signaling another entropic increase. The combined populations of Eloi and Morlocks is significantly less than the population at the end of the nineteenth century when Wells publishes *The Time Machine*. The absence of that energy relates the first and second law of thermodynamics. The necrotic decay of the species' biomatter infuses the environment with the energy for plant life that is further disordering the revenant architecture. The energy of the biosphere remains constant while the disintegration increases disorder. The components are then reordered by organisms that themselves speciate. The architecture additionally disintegrates by way of abiotic factors like weather, temperature, oxidization, and gravity. This process must also be unfolding across the planet. As a seat of globalization, London unlikely arrived at such disorder alone, and similar speciation and reclamation likely occurs elsewhere on Earth. Even if capital is the underlying schism, the global species is unlikely to evolve identical to the Eloi and Morlocks as the regional environments are quite diverse. Such evolutionary diversity is established by Darwin and thus well understood in the London imagination with which *The Time Machine* circulates. Entropy, as natural process, is expressed in the biodiversity and decay witnessed by the Time Traveller. Planetary disorder increases by natural, synonymously organic and physical, processes toward the maximum entropy equilibrium allows by the initial energy.[5]

The increase of entropy in the Earth system is inextricable from the disorder of the universal ecosystem, in reality and in *The Time Machine*. Reading Earth as a microsystem of the solar macrosystem links the increased planetary entropy with the orbital disorder at the far end of the novel's timeline. In primordial symmetry, Wells maintains the regressive theme when the physical location of the time machine, once London, is on a beach. The Time Traveller encounters giant versions of butterflies and crabs in a palindromic parallel of prehistoric time (Wells 2001, 145–146). Eventually, the novel presents life on Earth as green slime covering rocks and a black tentacular creature thrashing in the water (Wells 2001, 147–148). There is something disordered to these organisms in the anatomical manner by which they spread by slime and tentacle. They seem to phenotypically embody entropy while their revenant status models it. The disorder of molecular disintegration of the earth leaves only the most primordial visages at the end of the planet and sun.

At this end, the climate is quite cold, as the Earth is tidally locked with the sun that is no longer radiating as much heat (Wells 2001, 145). The disorder of the Earth system dissipates into lifeless mass as the sun does the same. The Time Traveller notes that an inner planet, likely Mercury, is passing unusually close to Earth signaling that disorder is increasing in the solar system as planets fail to maintain their orbits. The implicit kinetic energy of planets remains. The shift of planetary orbits likely results in some celestial collision, which will distribute kinetic energy and increase disorder in the system. Entropy is increasing, even if energy is to be found elsewhere in the universe. Tracing entropy, as physical concept, reveals a singular process by which the grandness of evolution and of solar systems unfold by the same natural processes. The unifying notion is that it all unfolds in spacetime.

The hook of *The Time Machine* is decidedly mathematical, asserting that geometry is inherently flawed: "any real body . . . must have Length, Breadth, Thickness, and—Duration" (Wells 2001, 60). The Time Traveller further claims that *"There is no difference between Time and any of the three dimensions of Space except that our consciousness moves along it"* (Wells 2001, 60). This modern notion of the universe was merely a theoretical argument when the novel was published. One year before *The Time Machine* and included in the plot, Simon Newcomb presents to the New York Mathematical Society a conceptualization of fourth-dimensional geometry that simultaneously competes with H.G. Wells and the Time Traveller to realize the fourth dimension (Wells 2001, 61; Newcomb 1894). Ten years before *The Time Machine* is published, the first serious mention of time as the fourth dimension occurs, which Nicholas Ruddick argues was possibly written by Wells or a student in his circle at that time (Ruddick in Wells 2001, 217). These two markers bind *The Time Machine* to what we now casually understand as the four dimensions of spacetime.[6]

Beyond a description of energy, entropy describes the thermodynamic arrow of time. In *A Brief History of Time*, Hawking states that "The increase of disorder or entropy with time is one example of what is called an arrow of time, something that distinguishes the past from the future, giving a direction to time" (Hawking 1988, 149). One can think of the progression of entropy as a glass falling off a table, moving through spacetime, and shattering on the floor. In breaking, the glass has become less bound and rigid in form, much like the Eloi and Morlocks progressing from humanity. Reading the split of humans into two species as a product of entropy, they become simpler but more chaotic and are no longer forced into the rigid societal binary of capitalist/worker. Instead, they have fractured into two separate species, which has always been the nexus for analysis of *The Time Machine*. This new view does not displace the old reading but encourages examination of how social and structural environments sustain natural selection and create biotic identity. By such ecocritical applications of physics to literature, we are able to interrogate our imminent biotic relation to spacetime before we inhabit the universal ecosystem.

Further, biotic relation to the thermodynamic arrow of time, and thus spacetime, is yet another proof of the directionality of time. This psychological arrow of time establishes directionality by human agreement that we experience the universe as an increase in entropy (Hawking 1988, 149). We never see shattered glasses reform, and we experience now before the future. Humans speciate into Eloi and Morlocks, not combine to devolve into their prior iteration. Biotic identity in nature is biotic identity in spacetime at all scales. The cosmological arrow of time notes directionality by the measurable expansion of the universe (Hawking 1988, 149). The total abiotic system further describes the same abiotic spacetime as localized thermodynamics and biotic perception. In the universal ecosystem and the Astropocene, biotic relations are uniformly described by physical laws. The arrows of time exemplify this relation: the thermodynamic arrow is a localized and scalable abiotic expression of spacetime; the psychological arrow relates biotic identity and thermodynamic causality; the cosmological arrow relays this at the maximum scale of the universe. In the Astropocene, nature and spacetime become indistinguishable—or rather, we internalize that they always were. Concepts of humanity are increasingly mathematical and fleshy, and thinking in this manner is to think in the Astropocene.

THE NUMINOUS OR MATHEMATICAL AND FLESHY

The Astropocene is initiated by biotic relation to the universal ecosystem as it is indistinguishable from spacetime. The Time Traveller is appropriately nameless, for how is identity not bound up in our perspectival relation to spacetime? His mobility in time fundamentally alters that most human assertion that we are our names. New biotic perspective recirculates ontology. The arrows of time indicate an inexorable spatiotemporality by which the entire reality of the universe, the biotic relations of organisms and environment, and the most basic cognitive action of witnessing causality all occur. The Time Traveller's identity is no longer rooted in terran causality and fundamental temporality. The Time Traveller has a novel relationship to spacetime and is unrecognizable, inconceivable by any other metric. His identity is subsumed by novel relation to spacetime. Recall that Watney is the Martian and Kathree is a Moiran Spacer in her third epigenetic iteration. If I can time travel, about what else do you care? Such an alteration of one's biotic relationship to the universe axiomatically alters one's ontological and social manifestation.

There are limitations to what ecocriticism extracts from *The Time Machine* concerning humanity in the Astropocene. The Time Traveller was always too deft at his craft for his contemporaries: "Things that would have made the fame of a less clever man seemed tricks in his hands. It is a mistake to do things too easily" (Wells 2001, 69). The time machine functions by a piece of technology that "looks singularly askew . . . there is an odd twinkling appearance about this bar, as though it was in some way unreal" (Wells 2001, 65). The mechanism is fantastical and a bit unstable in the reciprocating ecocritical method. *The Time Machine* brings the analysis this far by the veracity of natural selection and thermodynamics, in addition to the casual proof of spacetime. Wells, too, makes things seem mere tricks. If there be such a thing, *The Time Machine* is the hard science fiction of the Victorian Era. Nevertheless, ecocriticism prefers images of real nature, real spacetime.

Reading the construction of the Machine and the climactic encounter in Carl Sagan's *Contact* as a more physically viable version of Wells's travel narrative, ecocriticism examines the nature of a spatiotemporal environment as a feeling of approaching the numinous or becoming mathematical and fleshy. Trading up, Carl Sagan greatly advances the hard SF imagination of spatiotemporal travel narratives and the associated vehicle in his novel. Sagan's vehicle, the Machine, is far more mechanically detailed. While some components of the Machine are high-concept hard SF and "required whole new technologies to construct" (Sagan 1985, 266), all are less evasive than Wells's unreal bar. These technological processes and the assembly of their products are encoded into a complex signal suddenly detected by Ellie Arroway at the SETI Institute. Decoding the message and building the Machine inspires a global economic and engineering system, as well as pervasive human trust in its extensive plans. The engineering technology is far beyond the human knowledge base and all people are equally ill-equipped to assert what the assembled craft does. Humans gather materials in unprecedented quantities and employ strange crafting techniques. One component is particularly unsettling.

The Machine requires two copies of a component that disturbs the distance between organs and machines, a strange incarnation of bioengineering and cyborgs. To construct these pieces

> a particularly intricate set of organic chemical reactions was specified and the resulting product was introduced into a swimming pool-sized mixture of formaldehyde and aqueous ammonia. The mass grew, differentiated, specialized, and then just sat there—exquisitely more complex than anything like it humans knew how to build. It had an intricately branched network of fine hollow tubes, through which perhaps some fluid was to circulate. It was colloidal, pulpy, dark red. It did not make copies of itself, but it was sufficiently biological to scare a great many people. They repeated the procedure and produced something apparently identical. How the end product could be significantly more complicated than the instructions that went into building it was a mystery. The organic mass squatted on its platform and did, so far as anyone could tell, nothing. It was to go inside the dodecahedron, just above and below the crew area. (Sagan 1985, 257)

This synthetic biology strangely circulates with the various technologies that constitute the Astropocene. The organic components refract the holistic notion of genetically engineering lifeforms for terraforming. *Seveneves* utilizes genetic data to develop entire organisms, presumably by genotypical cellular reproduction in a simulated womb. While these growth chambers and genetic modifications give organisms a sort of cyborg identity, the plants and animals are then left to natural selection. *Contact* instead develops a singular and complex organic component out of a disproportionately simple process. In that way, it is precisely like natural selection.

However, the sufficiently organic entity then does nothing. No cellular reproduction. No movement. The component was an act of synthetic biology with primordial tones. An organic object emerges chemically. Predictably and repeatably, although not understood by humans, the component arrives in its form by adding chemicals to a separate, massive pool of chemicals that are offensive to terran senses. The component is additionally disturbing as it is disembodied viscera. Somehow, an organ can be constructed without body. If the Message instructed the growth of an organism which was then slaughtered and had an organ repurposed, there would be a distinct moral position attached to the organism's life, but the unsettling action is the killing and appropriation, not the formation. The act would be unsettling, but not in the same manner because everything is "according to the plan, even if the plan is horrifying" (Nolan 2008).

Alternatively, if the organ was some act of bioengineering with its shape genetically encoded and grown within a technologically simulated body, no terror ensues. *Seveneves* is not terrifying in its creation of novel planetary life because the actions are seated within the terrestrial process and knowledge base. Further unsettling, the synthetic organ then acts like disembodied viscera. The component does not twitch or flop. It does not attach to anyone. No SF horror ensues. The component does nothing more than any other component of the Machine. This unsettles because complex organs do not emerge from a chemical vat reminiscent of some swampy concoction coalescing to form primordial life. The chemical combination is so refined that it somehow always produces the component (Sagan 1985, 257). This synthetic biology uses the primordial mechanism but skips steps to form singular components that are chemical and fleshy, promethean and numinous. The component unsettles because its construction subverts the processes by which terrans know the organic to be constructed. Historical knowledge of terran biogenesis and subsequent biotic process at all levels is subverted by this visceral machine. And yet, lacking agency, the Machine is no cyborg.

Assembled, its purpose, activation, and mechanisms replace focus on the component technologies. The organic component is unceremoniously installed in the vehicle. Its unsettling character seems not to increase the uncertainty elicited by the total construct. Although once the Machine is turned on it will run uninterrupted for at least twenty minutes, performing whatever function its design must imply but is lost on terrans beyond notions of travel or communication. The Machine is shaped like a dodecahedron, has five seats, no other facilities imply a short journey, and requires the air pumped out of its surroundings to attain "the highest vacuum ever established on Earth" (Sagan 1985, 323). As the last of three pieces is moved into place, an electromagnetic charge builds. The Message says that, when fully charged, the Machine will activate. Sitting in the Machine, Arroway wishes she "had a child . . . her last thought before the walls flickered and became transparent, and, it seemed, the Earth opened up and swallowed her" (Sagan 1985, 323). The event encapsulates the entangled scales of biotic relations in the universal ecosystem.

Beginning with terran biotic relations, the assembled Machine is, initially, functionally indistinguishable from a spaceship, sustaining its terran crew despite vacuous environs. Hardly familiar to most terrans, the technological action is still well within the Anthropocene capabilities of the species. The vacuous environs are contained within the earthly environs. Terrans, at the behest of a far more universally entrenched species, have invoked within the planetary system the vacuous environs without. The layered disruption of internal/external binaries manifests biotic relation to the universal ecosystem. The weight of the circumstances inspires in at least the novel's protagonist a desire to be a parent—noteworthy in the final chapter this book, concerned as it is with biotic and cultural futures. The thought occurs at the exhilarating and terrifying locus of terran biotic relations to the universal ecosystem. The thought is lost as the Machine enters a wormhole.

Interacting with a wormhole is the maximum plausible relation with spacetime we have recognized. A wormhole functions by warping spacetime, which gravity is already doing. The traditional analogy is to imagine the universal spacetime as a sheet of paper with Points A and B drawn at opposite ends of the sheet. The theoretical physicist doing the explaining says that the points are too distant to reasonably travel by conventional means. She proceeds to imagine an alternative connection by bending the paper universe so the two points touch. This is often emphasized by jamming a pencil through the sheet, creating the wormhole. The two points are now connected and theoretically traversable. Theoretically, any point in space or time can be linked to another by a wormhole. The Machine enters such a wormhole, bringing Arroway and the rest of the crew to meet the beings who engineered the craft.

They emerge at the center of the Milky Way galaxy and gaze upon two black holes orbiting each other. The center of the galaxy acts as locus, as a hub, that connects the myriad wormholes snaking between the galaxy's star systems. Arroway describes the beings who brought them here as "those who make the numinous" (Sagan 1985, 367). These numinous beings have a near inconceivable control over their relation to the spatiotemporal environment of the universal ecosystem. Their chemical construction of an organic component for a craft that can sustain itself and organisms in a wormhole describes a species deeply and thoroughly understanding biotic relations to spacetime. These beings, whether they exist or are what we become, are at the climax of the Astropocene. The otherworldly beings tell Arroway that they, too, witness the numinous. They did not create the wormholes. They found them (Sagan 1985, 365–366). Some preceding beings created the revenant spatiotemporal architecture. Those who designed the Machine see evidence of these beings in transcendental numbers—communicating within unflinching mathematical truths that describe the patterned relations of numbers. Arroway is told to imagine that she found a message written in the eternal digits of pi (Sagan 1985, 367–368). The Anthropocene ends when we biotically relate to the universal ecosystem. And so begins the Astropocene, the spatiotemporal epoch characterized by ecometonymies best described by physics. Sagan may have done me one better here. In the numinous beings, we might sense a late-stage Astropocene, in which beings are about to control their interrelations to the universal ecosystems, linking their biotic identity to more abstract mathematical systems that would enable construction of wormholes. This is the nature of the Astropocene.

ECOCRITICISM FOR THE ASTROPOCENE

Biotically interrelated to spacetime, the Astropocene is mathematical and organic. I recall VanderMeer's literal *worm*hole in *The Southern Reach Trilogy* enmeshing the ecologies and star systems of two planets. I think of Watney's potatoes waiting on the Martian surface, produced by transient bacteria that are implied as much by their human's existence as Earth's gravity. I picture the thumb drive in *Seveneves* that carried the genetic library of total terran life, and I think about how the organisms were spawned from that data and then just left to evolve into whatever works. And I cannot help but to then think about the Purpose. There is something about thinking as nature in these images. There is something mathematically material and abstractly fleshy in ecocriticism for the Astropocene.

Ecocriticism is both an examination of human relation to nature and an ethic idealized by the likely unachievable prospect of speaking as nature. Earlier, entropy was traced along interdisciplinary connections that showed ecocriticism's access to biotic relations and evolution, planets and the universal ecosystem, and spacetime itself. Entropy describes biotic relations that define humanity in the universal ecosystem as a relation to spacetime itself. In pursuing human relation to nature, ecocritics must invoke and reconcile physics. Hawking assures us that general relativity and quantum mechanics are "the great intellectual achievements of the first half of [the twentieth] century" (Hawking 1988, 19) and that discovery of the quantum theory of gravity that resolves these mutually exclusive descriptions of the universal ecosystem is imminent. Perhaps we already have it and are just waiting for the technology to confirm the theory. Quantum mechanics and ecocriticism is a very different book from the spatiotemporal notions of nature thus far—although the topics do meet at blackholes, natural formations where things get extremely massive and singularly small. For now, general relativity and its descriptions of the universe are the focus.

The physics of spacetime is largely the purview of general relativity and subsequent derivations in theoretical physics. Einstein's theories describe the very massive objects in the universe and their gravitational impact on spacetime. General relativity predicts both the spatiotemporal fabric of reality and gravity's ability to warp it, while calculations based on Einstein's equations even predict the plausibility of wormholes existing in the universe, originally Einstein-Rosen bridges (Hawking 1988, 164). Published in 1915, general relativity has spent over a century shaping scientific and cultural relations to spacetime. General relativity has led to myriad explanations of the contours of the universal ecosystem, while literature and film have deployed its frameworks to explore humanness. Almost 100 years to the day after general relativity's publication, the predicted effects of gravity on spacetime are experimentally confirmed.

In late 2016, the Laser Interoferometer Gravitational-wave Observatory detects ripples in spacetime emanating out from the collision of two black holes (Abbot et al. 2016). LIGO maintains a laser whose beam is split into two perpendicular tunnels four kilometers in length and are reflected back to a detector. The wavelengths of the beams normally cancel each other out as they meet back at the detector. A significant gravitational wave passing through spacetime warps the tunnels separately and the beams do not cancel out upon returning, thus registering the gravitational wave rippling through the universe, through spacetime (Castelvecchi and Witze 2016). LIGO confirms the existence and impact of gravitational waves predicted by Einstein's theory of general relativity. The spatiotemporal fabric of the universal ecosystem is warped by gravity, as predicted, and ecocriticism has a new image of nature to consider.

I am not the first to consider general relativity's description of the universe as literary theory. In "Reading for the Wormholes: Micro-periods from the Future," Linda Charnes asks, "Is it possible to create a transmissible methodology out of reading for temporal 'wormholes'?" (Charnes 2007, 2). She considers sites alternative to cultural perspectives of time and conceptualizes literary moments, termed "wormholes." At these sites the future seems to appear in the present in an attempt to develop readings that recognize interpretations that are supposedly blocked by historicism's method of relating the past and present. The article attempts to run "time in the opposite direction, to see if it is possible to detect the future in texts" (Charnes 2007, 2). The article is engaging as a site of the paradigmatic circulation of general relativity in culture although its metaphorical use of wormholes manifests as anachronistic thought that is less elegant than or already deploying the historical logic of linear, unidirectional time.

Ecocriticism's diligent attention to scientific accuracy inoculates the literary theory against the metaphorical slippage at the core of "Reading for the Wormholes." Ecocriticism does wonder what human relations to nature become when the environment is characterized as flexible spacetime rippling with gravitational waves, when the detection of those waves verifies the model of the universe that also describes the viability of wormholes. Ecocriticism's axiomatic recognition of actual natural systems identifies the metaphorical slippage in the article. "Wormhole" is used to draw a line from present to the past and does not actually require wormholes to exist but rather uses a loose concept to justify the act. The underlying problem is that wormholes did not bend spacetime at any of the sites Charnes uses to exemplify literary wormholes. The assertion of these as wormholes invokes a nonexistent gravitas, that of physical calculations and dimensional reality, to cultural intersections that function by the causal logic of history and culture rather than the acausality so often associated with wormholes, the future altering the present.

Primarily, Charnes performs readings of the cultural perspective of time and history displayed in the Aymara language and a reportedly anachronistic editorial decision relating to Shakespeare's Falstaff. At the first site, Charnes details the unique cultural relation to time expressed in the language of the Aymara-speaking Andean tribe. The language and temporal perspective are undeniably fascinating, but Charnes's argument metaphorically conflates concepts of history and memory with the temporal dimension, as well as the biotic and anatomic with the cultural. The Aymara language perceives the past in front of a person, as it is conceptually available while the future is at their back, out of sight. This manifests as gesturing behind when speaking of future events and signals an alternative approach to history, all laid out in front (Charnes 2007, 2). The Núñez and Sweetser paper invoked by Charnes examines these "temporal metaphor systems" (2006, 401), deftly describing the exclusively cognitive experience of time expressed in language and gestures. The Aymara language constructs temporal metaphors—not biotic relations to spacetime—in which the speaker is cognitively facing the events of the past. Hardly a biotic relation to time, such applications of wormholes undermine the ecocritical assessment of spacetime necessary for perceiving the Astropocene. Any philosophy derived from perspectives of history and memory is barely related to the dimension of time. Here, "Reading for the Wormholes" problematically conflates spatiotemporal causality with the memory and history that construct a social present.

Interestingly, Charnes approaches a biotic relation to time in the reading, only to once again be foiled by the metaphoric. Charnes offers that the "alignment of past and future with the placement of the eyes in one's head means that the Aymara literally *look at, or see*, time differently than we do" (Charnes 2007, 2). Here, Charnes seems to dissolve her previous distinction between temporality as the politic of history and the dimension of time which "has nothing to do with us" (Charnes 2007, 2). Ecocriticism takes issue with Charnes's use of "literally" (2007, 2) when describing the future behind the Aymaran people, as the notion is conceptual and metaphorical, not actual. The metaphorical slippage makes things problematically indistinct. This is not ecometonymy or ecorealism in the universal ecosystem. I am inclined to think that those who speak Aymara can recognize the thermodynamic arrow of time. Why would figuratively aligning vision with memory of the past convert perceived reality such that a tree decaying is at the beginning of its life. They are not a group of Merlyns, saddened by hellos that signal the last time they will ever see someone (White 1939, 29). What Charnes calls a wormhole is merely a recognition that the present contours our relation to the past. Perhaps this is more foregrounded for speakers of Aymara, but it is not a wormhole, not a biotic relation to spacetime. Charnes is derailed by ratio-nalizing this as a wormhole that moves present influences into past events.

Charnes figuratively adorns this problematic anachronism with the physi-cal wormhole to assess an impactful bit of editing in Shakespeare's *Henry V*. More Charnes's field than mine, I offer only a brief summary of the events that constitute the supposed wormhole. Shakespeare's folio text is corrupted and a line is corrected. The correction is continually reinforced by Shakespearean scholars and "makes so much sense as to go relatively unchal-lenged to this day [revealing] less about Fallstaff than it does about how his character has been read by successive centuries" (Charnes 2007, 5). Charnes provides fascinating context for the emendation by reading the shift in nostal-gia as a disease of the mind first defined spatially and later characterized as a longing for an idealized time. As far as I can tell, the reading is a wonderful literary history of *Henry V* and Fallstaff manifesting in material culture.

Alternatively, Charnes reads this as an assertion of the nostalgic con-dition and its imperial connections, which are not established until after Shakespeare. By its insertion and general acceptance, Charnes reads a worm-hole through which the future erupts into the past. There is a fascinating recursion, and recognition of the edit could theoretically disappear, making it as though they were the original words and that Shakespeare prophesied nos-talgia. This is not a wormhole though. A wormhole implies physical travel in spacetime, such as someone traveling back to create the corruption, or filling it in such that it was never known to not be the bard himself. Only when his-toricized does the wormhole become apparent, and by historicizing the event, a more logical and elegant notion appears: historicism does its normal work and notes a cool editorial moment. The wormhole is extinguished. Rather, the idea of the wormhole is extinguished for it is not at all the process by which the events unfold.

I am not opposed to "Reading for the Wormholes: Micro-periods from the Future" as a text itself, where the writer earnestly attempts to read in a post-general relativity paradigm, to think with spacetime and wormholes. I am similarly intrigued by the Aymara notions and see a potential path to include their spatial approach to temporal cognition in a discussion of nontraditional figurings of spacetime emerging separate from the general relativity paradigm. I wonder what literary criticism and history becomes if we ever find, utilize, or create an actual wormhole. To warp spacetime such that one arrives elsewhere in the universe or time alters what it means to be terran and establishes a new biotic relationship to the universe. Utilizing wormholes is the type of thing that could end the Astropocene. Perhaps then we will find a method to read with wormholes. Using a wormhole could be the last astropic act, a bookend to Watney's use of the Martian atmosphere to preserve his earthly circle of nature. On the cusp of the late-stage Anthropocene though, we must sufficiently imagine biotic relations to the universal landscape to make ecoethical progress toward the Astropocene.

Perhaps gravitational waves are an example of what Charnes was going for. Consider the anachronistic action in the cultural paradigm extending from the 1915 publication of general relativity along the continuum to whatever the Astropocene reveals about our biotic relation to spacetime. Perhaps the 2016 confirmation of gravitational waves acts more like the historical wormhole Charnes means. After 100 years of working as though gravitational waves ripple through spacetime, an experiment confirms it. The finding is principally as valuable as the theory, a monumental affirmation and maintenance of Einstein's work and the scientific method itself. But after a century of other forms of validation, the authorization of general relativity by the experiment does not issue a paradigm in our temporal location but rather affirms the paradigm that general relativity already issued. There is a sort of reciprocating action to this chronology that additionally synonymizes the supersystems of nature and spacetime.

Other texts help conceptualize the paradigmatic thought post-general relativity that binds humanness with spacetime—albeit at the expense of science fact. This close to the final pages, after ecocritically reciprocating science fact fiction to examine the end of the Anthropocene, I am more comfortable invoking the feel of reading in the Astropocene. For unprecedented perspectives on spacetime, I turn to Kurt Vonnegut's *Slaughterhouse-Five* (1969) in which Billy Pilgrim becomes unstuck in time, living his life in a nonlinear progression. Billy lives his death, birth, and everything else in a chaotic order. Additionally, he is abducted by the Tralfamadorians. These beings offer the novel an additional relationship to temporality: they can see all of time for any being. From this vantage, they have identified several more than just the two sexes needed to continue the human species (Vonnegut 1969, 145–146). *Slaughterhouse-Five* is set in World War II, although during the Berlin bombing, so it only maintains adjacent associations with the Anthropocene's atomic character, but seeing tendrils extending to the spatiotemporal landscape of general relativity is hardly a conceptual misstep. Alternatively, Dr. Manhattan of Alan Moore and Dave Gibbon's *Watchmen* (1986) is deliberately entrenched in ominous associations with atomic bombs. He was disintegrated when bombarded with tachyons, a fictional particle that supposedly expresses time. Now he is an all-powerful being simultaneously experiencing every moment of his infinite life. *Arkham Asylum* (1989), a graphic novel set well before Bruce Wayne becomes Batman, extends into the past the axiom that the very existence of Batman conjures into being his villainous foils. Before the existence of Batman, he is erupting into the past to create his future.

Arrival (2016) by Denis Villeneuve actually does what Charnes wants the Aymaran language to do. Aliens arrive and grant humans their semasiographic written language which fundamentally expresses their cognitive and biotic relation to time. The main character becomes fluent and witnesses the future by the principles that language constructs perspective. The aliens give the language, the ability to see the future, to humans in the hopes that it will save the future. A lighthearted counterpoint to the exceptional and intense *Arrival*, an episode of *The Big Bang Theory*, "The Focus Attenuation" investigates the grammar of time travel for discussing *Back to the Future*: "Is 'placed' the right tense for something that would have happened in the future of a past that was affected by something from the future?" The characters decide on "had will have placed" as the proper tense (Cendrowski 2015). *Avengers: Infinity War* (2018) describes how taking control of the very fabric of reality is the way one might resolve the ecological problems of exponential population increase in a finite system. The sequel, *Avengers: Endgame* (2019), discards a discussion of composting because the technosavior figured out time travel—so they just use that. The film also takes the time to clarify the actual structure of time, as opposed to that propagated by time travel movies like *Back to the Future* (1985). These are all iterations of post-general relativity perspectives circulating in our culture. The Astropocene makes nature spatiotemporal and increasingly experienced by total terrans.

This train of thought prepares ecocriticism for the Astropocene, focusing on its ability to reflect on biotic terrans in spatiotemporal nature, and thus shape the cultural expression of that identity for the Astropocene—the above texts evidence that we are already entrenched in spatiotemporal construction of environment. The ultimate problem of Charnes's article is the inability to reconcile science fact and fiction to produce a cultural model.

This is ecocriticism's next task. Gargantuan, but "the civilizations with only short-term perspectives just aren't around" (Sagan 1985, 361). Sagan numinously describes the very threat of climate, conflict, and extinction in the Anthropocene. Biotic relations are mathematical and fleshy now, and biotic identity is soon to be contoured by interrelation with the universal ecosystem. Ecocritical methodology circulates the information necessary to foresee and curate the Astropocene. Ecocriticism can be the next paradigmatic perspective, ending this epoch and shaping the next.

NUMINOUS BEINGS ARE WE

It has been over twenty years since E. O. Wilson gave us our charge: "The greatest enterprise of the mind has always been and will always be the attempted linkage of the sciences and humanities" (Wilson 1998, 8). Wilson then initiates a consumptive disciplinary unification prioritizing biology. I offer only a slight, but critical shift: begin in environmental studies, specifically that transitional space between the environmental sciences and environmental humanities. In this interdisciplinary space, *The End of the Anthropocene* imbues ecocriticism with the ability to draw in literature and mathematics, and everything in between—sometimes just everything. In this way, ecocriticism can be a paradigmatic technology for understanding and shaping the next geological—but more importantly, terran—age.

Wilson asserts that every college student, public intellectual, and political leader should be capable of providing an answer to the question: "What is the relation between science and the humanities, and how is it important for human welfare?" (Wilson 13). E. O. Wilson offers "consilience" as his method for fleshing out the relationship between these seemingly disparate fields. This move is linguistic, choosing the term because "its rarity has preserved its precision" (Wilson 8). Wilson credits William Whewell with the definition, explaining consilience as "literally a 'jumping together' of knowledge by the linking of facts and fact-based theories across disciplines to create a common groundwork of explanation" (Wilson 1998, 8). For the larger part of his book, Wilson explores this groundwork and seeks a pathway from the sciences to the arts.

Wilson begins this movement by constructing a logical, perhaps even natural, path: determine the physical properties of the world around us, understand how they explain human social behavior, and use them to determine the role of the arts. *Consilience* also seeks to unite the intended outcomes of the sciences and humanities arguing that "There has never been a better time for collaboration between scientists and philosophers. . . . Philosophy, the contemplation of the unknown, is a shrinking dominion. We have the common goal of turning as much philosophy as possible into science" (Wilson 1998, 12). Here Wilson defines the goal of consilience as the proper action of any thinking individual.

In this way, Wilson also examines the methods of history. He argues that "History is today a fundamental branch of learning in its own right, down to the finest detail. But if ten-thousand humanoid histories could be traced on ten thousand Earthlike planets, and from a comparative study of those histories empirical tests and principles evolved, historiography . . . would already be a natural science" (Wilson 1998, 11). For Wilson, intellectual pursuit is innately scientific, seemingly limited only by observable sample size. Wilson sees consilience as the next great scientific framework for the unification of all knowledge and constructs a version of interdisciplinarity that universally employs the scientific method as a way of accessing and explaining all things, moving from the physical laws of the universe to the neurological components of art and the biological imperative of storytelling. I am less enthusiastic about such a straight line.

For a model of consilience, I prefer swamps. As a child, I had dedicated frog-catching clothes, because if there was muck and water, I was in it and routing out frogs, snakes, and turtles. The best my mom could hope for was that I would change first. Years later, I still find comfort moving through muck, and I have crawled and swam in many swamps. One of the most beautiful places I have been is a bit of swamp on the East Coast where the tannins leached from the cedars imbue the water with a red tint. Swamps are transitional landscapes of water and vegetation teeming with biodiversity. In some ways, they are primordially generative and in others they are fertile seats of humanity. There is a special place in my heart for the difficulty they present in land development. Elsewhere, I have ruminated upon "the nature of the swamp as an integrative biome cyclically blurring life with decay, 'a place where wild process dominates' [Snyder 2004, 5]; the muck and moisture break down dead organisms quickly while releasing nutrients back into the ecosystem" (Gormley 2019, 38). I would take the model of the swamp, teeming with life and rife with biotic cycling, inaccessible and fertile, and slide the swampy framework into consilient connectivity.

Mostly, I find myself in swamps to track. The water and biodiversity make swampy spaces a consistent inclusion in the daily business of a wide array of species, and swamps' transitional identity brings into contact the terrans of the adjacent waterways and those at home in the areas of firmer ground. I have noticed animals tend to be more at ease in swampy areas. There is protection in the vegetative tangles interspersed with waterways that provide wide sightlines. The soft ground and intricate plantlife do not necessarily favor large predators running after a prey animal. Of course, I'm located in the northeast United States, so my local fauna are less concerned with predators in the water. Most importantly, humans tend to avoid swamps. There are all sorts of tangled reasons that are personal, cultural, and infra-structural. Most of it seems to be an aversion to muck and the critters moving about. Swampy, soft ground though makes for good, clear tracks, and everything heads to the water at some point. The whole space helps trackers do their thing.

Early in this book, I invoked tracking as an ecocritical framework, drawing heavily on Louis Liebenberg, the mathematician who learned to track while living with the Kalahari Bushmen. In *Consilience*, Wilson also finds his way to Liebenberg as a site of interdisciplinary connectivity. Wilson's process of finding connection points from the sciences to the arts creates a structural motif of linearity that draws out of tracking the same singular directionality. In *Consilience*, tracking is entrenched in the pursuit of a single animal. The hunters find tracks and sign that allow them to connect to the next track. Wilson even addresses the speculative tracking of the Kalahari Bushmen in which they attempt to enter the mind of their quarry. Everything beelines for the next indicator, be it the tracking or the book. Tracking as parsing a linear path of connectivity is well suited for Wilson's methodology. Here, though, it is not swampy enough, and fails to recognize the nonlinear ecometonymic interrelations expressed in tracks.

Tracks are ecometonymies displaying the massive biotic mesh of interrelation between organisms and their environs. Tracing those connections, tracks became indistinguishable from the organism itself as it exists in an ecological space and time. The tracks describe an organism acting in the immediate environment and the organisms it encounters. The act of tracking invokes a strange spatiotemporal mindset as the track manifests in the tracker's present an image of the quarry's past. If the object is to pursue that being, then the tracker seeks to collapse her own and the quarry's present in an often biotic interaction. While hunting almost always implies tracking, tracking does not necessarily imply hunting and is an access point to biotic identity which, as used throughout this book, reveals the character of the Anthropocene and Astropocene. Doubled or not, in one's biotic home or in some extraplanetary environ, biotic identity is evidenced in the tracks. This chapter imagines the contours of biotic relations to spacetime, drawing throughlines to theoretical physics from ecocriticism's more traditional environmental access points. This analysis winds down to tracking as an access point to a swampy revision of interdisciplinarity that can be recirculated as an ecoethic for the end of the Anthropocene.

With an ecocritical view of the end of the Anthropocene, I offer two amendments to Wilson's approach as we end the Anthropocene and enter the Astropocene. The first is to begin not in biology but in environmental studies, specifically in the transitional site of the environmental sciences and environmental humanities. The imminence of the late-stage Anthropocene makes this choice simple. While these two areas of study function quite differently, I see many traversable paths between them. Ecocriticism's focus on narrative relations, scientific literacy, and assessment of actual nature seems a particularly useful methodology for interrelation. Second, replace linearity with the swampy and outward interrelations of tracking and spacetime. This book spends its time ecocritically entwining many disciplines and texts to recognize the contours of the Anthropocene that become the contours of Astropocene. I see no greater utility or perspective than that. Ecocriticism can recognize and thus impact the characteristic environmental control of the late-stage Anthropocene to then shape the Astropocene. Terrans need to see further if we are to do this much longer.

As this penultimate section deals with tracks, a sort of symmetry insists that I turn one last time to wolves.

WOLVES AT LAST

The Astropocene is now before us. We will shortly undo our doubling and again have a planet that does not refuse us. This action sets the global ecoethic that defines our interactions in the universal ecosystem. This book characterizes the Astropocene so we can include it in the ecoethical process that ends the Anthropocene. This ecoethic is our effort to enact that impossible task: speak as nature. So here in these last pages, I loosen the bounds and let slip science fact fiction into science fact fantasy to invoke tracking and biotic interrelations as the path by which we approach speaking as nature—which sometimes means letting our ecometonymies do it for us.

Science fact fantasy is plausible and chaotic. It builds fantastical worlds that abide scientific findings. Science fact fantasy is scientifically plausible, just statistically staggering. Science fact fantasy imagines a coalescence of matter and energy somewhere in the infinite and chaotic universal ecosystem that abides and progresses by natural law but does not wholly require it to be real. *The Martian* and the first two books of *Seveneves* are science fact fiction. The final section of *Seveneves* has a scale that approaches science fact fantasy, although Stephenson painstakingly clarifies the technological ability for humanity to create such a world and reduces the fantastical to its minimum. *Contact* is firmly science fact fantasy during the climactic travel narrative, the first contact event, and for much of the Machine's construction. Alternatively, M.H. Bonham's "The Company You Keep" is a distinctly science fact fantasy short story that coalesces the ecometonymic, biotic, and interplanetary throughlines of *The End of the Anthropocene*. Also, it has space wolves.

"The Company You Keep" is the penultimate inclusion in *A Kepler's Dozen*, a conceptual SF collection in which the authors compose short stories set on exoplanets discovered by the Kepler telescope. Each story is prefaced by astronomical data on the planet and its star provided by scientists "involved in the quest for exoplanets" (Howell and Summers 2013, 2). Bonham's story is set on a moon orbiting Kepler-16b. This snowy orb with views of the twin suns elicits the necessary *Star Wars* references by exobiologist Cinthia. The distance of the planet and moon from the twin suns creates the cold climate. Cinthia won the Nobel Prize for discovering the first nonterran lifeform and is on the moon to catalogue the wildlife. Her companion is a dog she was assigned by NASA. Originally Laika, Cinthia affectionately renames him Wolf. The story opens with the pair gathering a new critter they decide to call a Kepler ice crab. Cinthia bags the animal and heads back to the base to get it into stasis before it dies, implying experience with biotic incompatibility between the planet's organisms and the earthly facilities. More cynically, this may signal a less than integrative but characteristically human approach to exploration and study. Cinthia is interested in the being's survival and her relationship with Wolf positions her as the principle ecoethical agent of "The Company You Keep."

In addition to the ice crab, Cinthia notes that they have found an eight-legged, long furred, herbivorous herd animal that tends "to graze at night in the coldest darkest times making [her] wonder what hunted them" (Bonham 2013, 176–177). Cinthia remarks that they have seen quadrupedal omnivores, but their relatively small size precludes them from eating the horned beasts, except as carrion. When Wolf harries the herd, he is chased off when they charge him, "heads lowered with four horns on their heads and sharp spine ridges that served as armor" (Bonham 2013, 177). The horned beasts even have a sense for detecting even the slightest heat signature which, Cinthia says, "enables them to find liquid water and the plants they prefer" (Bonham 2013, 185). These traits—herbivorous, living as a herd, reacting to Wolf as a threat, evolution of offensive and defensive anatomy, feeding at night, sensing heat—all also imply an apex predator evolved and sustained by a large protein source. Cinthia has yet to find one, but her ability to read the ecometonymic biotic implications of the horned beasts has her "looking for a top-tier predator" since they arrived (Bonham 2013, 181).

Reading tracks, Cinthia ecometonymically constitutes physical traits of the apex predators—otherwise only implied by herds of large fauna—adapted to a snow-covered moon at the far limit of its solar system's habitable zone. The group discovers that Kevin, Cinthia's junior partner, was in a vehicle wreck while gathering samples alone. The metal vehicle is scratched and there are tracks: "Paw marks in the snow. Lots of them with claw indentations. Six-toed marks with generous pads, splayed to keep traction on the snow and ice" (Bonham 2013, 181). While very canid looking—the presence of nonretractable claws is the simplest distinction between canid and feline tracks—these are extraplanetary organisms produced in a non-earthly biotic context. (This is science fact fantasy, completely plausible and never guaranteed.) The tracks and sign show a likely aggressive animal suited for deft maneuvering in the frozen environment intelligently investigating the vehicle. Cinthia corrects an assumption of their size based on the track size, considering an ecometonymic expression of environment: "the feet could snowshoe out for traction [and] greater stability in the snow" (Bonham 2013, 182). The large feet may not scale as expected. A nimble tracker, Cinthia distinguishes at least four separate organisms investigating the vehicle. The search party sees Kevin's tracks leave the scene. The wolfish tracks follow, overlaying the terran's. They say it is hard to tell, but Kevin seems to be running.

As Cinthia and Wolf follow the tracks into the forest, human-canine boundaries dissipate, and an interplanetary confrontation initiates a model where earthly kin-making makes kin elsewhere. Cinthia keeps Wolf on a leash so he does not bound off and get separated or dangerously entangled with the Kepler wolves. Wolf is still scent tracking Kevin although the tracks are clear to Cinthia as well. The archetypal military guy has his gun out and is using a long-range sensor to look for the Kepler wolves' heat signatures, which is not registering anything—remember that they hunt animals who have an organic version of this ability. He switches the sensor to "focus on CO_2 emissions" (Bonham 2013, 185), and the group discovers they are surrounded. Wolf breaks free of the leash and confronts one of the Kepler wolf pack. The two begin to play—licking, pawing, and bounding with each other. The rest of the pack investigates the other terrans and are quickly accepting. Wolf vouched for them. Kevin climbs down from a tree and the terrans are included in the Kepler wolf pack.

Wolf is an ancient technology. Inextricable from each other's histories, wolves, dogs, and humans circulate with the very core of the Anthropocene. The dog, whether a domestication of wolves or a parallel emergence of some common prior species, is as much an ecometonymy of humans as wolves. Wolf is an incredibly transformative ancient bioengineering project that is ultimately about interrelating with the wilds. Cinthia embodies this. Her dog is not some historic space animal—laika is just Russian for dog—but an historic terran. Wolf facilitates human connectivity with space wolves, giving humans the chance to reissue relationships to apex predators that must be critical to maintaining the ecosystem of Kepler's moon. This may be the very same chance we have with wolf reintroduction projects.[7] Demonstrating a critical ecoethic for the Anthropo- and Astropocene, Wolf and Cinthia are companion species making more interspecies companionship.[8] This is an easy framework for making companion species, and is an ideal(ized) image for an ecoethic that brings us to the end of the Anthropocene. Anyways, the human genetic history, ecometonymically if not literally, already contains something about dogs and wolves.

"The Company You Keep" posits an ecoethic with a bit more depth than implied by its delightfully campy titular one liner. Bonham describes reciprocal interspecies maintenance of each other in the universal ecosystem. Beyond her relationship to Wolf, Cinthia even wishes she had wool clothes instead of the high-tech heat suit that leaves her sweaty and pruned. She would prefer natural fibers helping her body regulate temperature in relation to the environment, rather than a suit that resists the environs. "The Company You Keep" even goes as far as identifying the Kepler wolves by their CO_2 emissions, hilariously exemplifying the biotically appropriate level of greenhouse gas production. Cinthia did not need such technology to know they were being followed, stating that she could sense it, like so many beings when integrated with their environs. When facing a pack of wolves, she lets her dog speak for her, and humans interrelate with pristine wilds. In the opening of *The Crossing*, Billy too is gazed upon by a pack of wolves—those great biotic wardens who perpetuate their prey by hunting their prey—then the wolves walk away from him and, in many ways, the world itself. Billy never tells anyone. *The End of the Anthropocene* is a view of both options. This is where I leave you.

We are about to decide how anthropos interacts with planetary nature, shifting to control our impact on planetary ecology. Whatever this action, it quickly becomes our permanent presence in the universal ecosystem. The Astropocene entices and exhilarates because it happens out in the wilds of spacetime. The epoch is largely shaped by maintaining our biotic identity as we interrelate with the universal ecosystem. By ecocriticism, we now have information from the Astropocene with which we can true ourselves in the present. We are fast approaching the moment where we speak as nature or instead of nature. Along the second path, we ecologically control the Earth system and form a planetary ecology of anthropos. Likely, the resulting ecosystems will unfold in their chaotic way, and we will occasionally or incessantly reissue the parameters for anthropogenic nature. The other path is maintenance rooted in biotic terran relations that promote wild processes. Speaking as nature is often just letting nature unfold. This is our last chance, for a long while anyway, to develop an ecoethic of maintenance by biotic interrelation. We must be as wolves, great biotic wardens whose wild acts invigorate each terran being.

NOTES

1. At the theoretical temperature of absolute zero, a particle reaches its minimum energy and entropy. Approaching such temperatures, atoms exhibit strange behaviors. Some atoms become superconductors whose electrical resistance drops to zero

and can circulate a current indefinitely with no power source. Others become Bose-Einstein condensates, the fifth state of matter. Absolute zero bumps into quantum mechanics when all energy is dissipated and concepts of zero-point energy emerge.

2. *The Time Machine*, a product of Victorian London, has plenty of whiteness in its imagery. Perhaps the self-consumptive devolution extending from the seat of the industrial West can act as a sort of reprieve.

3. *On the Origin of Species* is published in 1859, *The Time Machine* in 1895.

4. Here, one might take issue with the small timescale on which the species diverges. I find the year being 802,701 a rather cosmetic detail that was procured, by the way, by Wells's devotion to cutting-edge scientific theory. It is safe to extend Wells's scientific diligence to the time needed to alter the species as a handicap of the age's knowledge rather than earnest attempts at veracity. All of that aside, I think it is rather simple to pretend it said a better number or that the number was better, and the book wouldn't be very sci-fi without an exact calculation. If that doesn't work, let's recall that in *Seveneves* our species split three ways in just 5,000 years. One group of which used selective breeding.

5. The planet is, of course, not a closed system so it does not have to increase in entropy immediately, but the universe is a closed system increasing in entropy, so there is little gained by lingering on a model of *The Time Machine* where entropy is decreasing on Earth while increasing elsewhere in the universal macrosystem. The planet and its inhabitants are no less developing in the universal ecosystem. The entropy of the solar system, as a microsystem of the universe, is certainly increasing.

6. An excerpt from Newcomb's lecture is included in the unimpeachable Broadview Edition edited by Nicholas Ruddick.

7. I would note that M. H. Bonham is a Montana resident. Wolf talk always has a bit more exigence emerging from a state that houses Yellowstone.

8. Which, by the way, shares an etymological root of "breaking bread" which implies a sharing that runs counter to the capitalist iteration, "company."

Bibliography

Abbot, B.P., LIGO Scientific Collaboration, and Virgo Collaboration. "Observation of gravitational waves from a binary black hole merger." *Physical Review Letters* 116, no. 061102 (2016), 1–16.

Abrams, J.J. *The Force Awakens*. San Francisco, CA: Lucasfilm, 2015.

Alston, J.M., B.M. Maitland, B.T. Brito, S. Esmaeili, A.T. Ford, B. Hays, B.R. Jesmer, F.J. Molina, and J.R. Goheen. "Reciprocity in restoration ecology: When might large carnivore reintroduction restore ecosystems?" *Biological Conservation* 234 (2019), 82–89.

Alvarez, Luis W., Walter Alvarez, Frank Asaro, and Helen V. Michel. "Extraterrestrial cause for the cretaceous-tertiary extinction." *Science* 208, no. 4448 (1980), 1095–1108.

Anger, Suzy. "Evolution and entropy." In *A Companion to British Literature Volume IV: Victorian and Twentieth-Century Literature 1837–2000*, edited by R. DeMaria, H. Chang, and S. Zacher. Malden: Wiley Blackwell, 2014. doi: 10.1002/9781118827338.ch79.

Armstrong, Robin. "Time to face the music: Musical colonization and appropriation in Disney's Moana." *Social Sciences Special Issue: The Psychological Implications of Disney Movies* 7, no. 7 (2018), 1–9.

Baldacci, S., S. Cerrai, G. Sarno, N. Baïz, M. Simoni, I. Annesi-Maesano, and G. Viegi. "Allergy and asthma: Effects of the exposure to particulate matter and biological allergens." *Respiratory Medicine* 109, no. 9 (2015), 1089–1104.

Barrie, David. *Sextant: A Young Man's Daring Sea Voyage and the Men Who Mapped the World's Oceans*. New York: William Morrow, 2014.

Berry, R. Stephen. *Three Laws of Nature: A Little Book on Thermodynamics*. New Haven: Yale University Press, 2019.

Beschta, Robert L., Luke E. Painter, and William J. Ripple. "Trophic cascades at multiple spatial scales shape recovery of young aspen in Yellowstone." *Forest Ecology and Management* 413 (2018), 62–69. doi: 10.1016/j.foreco.2018.01.055.

———. "Trophic Cascades and Yellowstone's aspen: A reply to Fleming (2019)." *Forest Ecology and Management*. doi: 10.1016/j.foreco.2019.05.014.

Beschta, Robert L., and William J. Ripple. "Can large carnivores change streams via a trophic cascade?" *Ecohydrology* 12, no. 1 (2018), 1–13. doi: 10.1002/eco.2048.

———. "Riparian vegetation recovery in Yellowstone: The first two decades after wolf reintroduction." *Biological Conservation* 198 (2016), 93–103. doi: 10.1016/j.biocon.2016.03.031.

Biesalski, Hans K. "Nutrition meets the microbiome: Micronutrients and the microbiota." *Annals of the New York Academy of Sciences* 1372 (2016), 53–64.

Blakeslee, Nate. *American Wolf: A True Story of Survival and Obsession in the West.* New York: Penguin Random House, 2017.

Bonham, M.H. "The Company You Keep." In *A Kepler's Dozen: Thirteen Stories About Distant World's that Really Exist*, edited by Steve B. Howell and David Lee Summers. Mesilla Park, NM: Hadrosaur Productions, 2013, 175–187.

Brown, Tom, Jr. *The Science and Art of Tracking: Nature's Path to Spiritual Recovery*. New York: Berkley Books, 1999.

Brown, Tom, Jr., and Brandt Morgan. *Tom Brown's Field Guide to Nature Observation and Tracking*. New York: Berkley Books, 1983.

Buell, Lawrence. *The Future of Environmental Criticism: Environmental Crisis and Literary Imagination*. Malden: Blackwell Publishing, 2005.

Carey, John. "Are we in the Anthropocene?" *PNAS* 113, no. 15 (2016), 3908–3909.

Castelvecchi, David, and Alexandra Witze "Einstein's gravitational waves found at last: LIGO 'hears' space-time ripples produced by black-hole collision." *Nature*. Accessed 28 August 2020. https://www.nature.com/news/einstein-s-gravitational-waves-found-at-last-1.19361.

Charnes, Linda. "Reading for the Wormholes: Micro-periods from the future." *Early Modern Culture*, no. 6 (2007). Accessed via Google cache 12 September 2017. http://emcserver.org/1-6/charnes.html.

Cheshire, James, and Oliver Uberti. *Where the Animals Go: Tracking Wildlife with Technology in 50 Maps and Graphics*. New York: W. W. Norton & Company, 2016.

Clark, Timothy. *Ecocriticism on the Edge*. New York: Bloomsbury Academic, 2018.

Clements, Ron, and John Musker. *Moana*. Burbank: Disney, 2016.

Crutzen, Paul J., and Eugene F. Stoermer. "The 'anthropocene.'" *IGBP Newsletter*, no. 41, May (2000).

Cuaron, Alfonso. *Gravity*. Burbank, CA: Warner Home Video, 2013.

Curnow, Joe, and Anjali Helferty. "Contradictions of solidarity: Whiteness, settler coloniality, and the mainstream environmental movement." *Environment and Society* 9 (2018), 145–163. https://www.jstor.org/stable/26879583.

Darwin, Charles. *On the Origin of Species by Means of Natural Selection*. Rpt. from 1859. Edited by Joseph Carroll. Ontario: Broadview Press, 2003.

Derrida, Jacques. *H.C. for Life, That Is to Say . . .* Translated by Laurent Milesi and Stefan Herbrechter. Stanford, CA: Stanford University Press, 2006.

———. *Of Grammatology*. Translated by Gayatri Spivak. Baltimore, MD: John's Hopkins University Press, 2016.

————. "Plato's pharmacy." In: *Dissemination*, translated by Barbara Johnson. London: The Athlone Press, 1981.

Einstein, Albert. "Autobiographical notes." In: *Albert Einstein, Philosopher-Scientist: The Library of Living Philosophers Volume VII*. Peru: Library of Living Philosophers, 1969.

Ellis, Erle C. *Anthropocene: A Very Short Introduction*. Oxford: Oxford University Press, 2018.

Fackrell, Laura E., Paul A. Schroeder, Aaron Thompson, Karen Stockstill-Cahill, and Charles A. Hibbitts. "Development of martian regolith and bedrock simulants: Potential and limitations of martian regolith as an in-situ resource." *Icarus* 354, no. 15 (2021), 1–14.

Ferrando, Francesca. "Posthumanism, transhumanism, antihumanism, metahumanism, and new materialisms: Differences and relations." *Existenz* 8, no. 2 (1999), 26–32.

Filoni, Dave. *Star Wars: The Clone Wars*. United States: Warner Bros, 2008–2020.

Fleming, P.J.S. "They may be right, but Beschta et al. (2018) give no strong evidence that 'trophic cascades shape recovery of young aspen in Yellowstone National Park': A fundamental critique of methods." *Forest Ecology and Management*. doi: 10.1016/j.foreco.2019.04.011.

Foa'i, Opetaia, and Lin-Manuel Miranda. "We Know the Way" Track 5 on *Moana: Original Motion Picture Soundtrack*. Walt Disney, 2016.

Fogg, Martyn. "Planetary engineering bibliography." Last modified January 2011. Accessed 7 March 2021. http://www.users.globalnet.co.uk/~mfogg/biblio.htm.

————. *Terraforming: Engineering Planetary Environments*. Warrendale: Society of Automotive Engineers, 1995.

Francis, Matthew R. "The hidden connections between darwin and the physicist who championed entropy." *Smithsonian Mag*, 15 December 2016. Accessed 13 July 2020. https://www.smithsonianmag.com/science-nature/hidden-connections-between-darwin-and-physicist-who-championed-entropy-2-180961461.

Francis, Richard C. *Epigenetics: How the Environment Shapes Our Genes*. New York: W.W. Norton and Company, 2011.

Garrett-Bakelman, Francine E., et al. "The NASA twins study: A multidimensional analysis of a year-long human spaceflight" *Science* 364, no. 144 (April 2019), 1–20. doi: 10.1126/ science.aau8650.

Gerstell, M.F., J.S. Francisco, Y.L. Yung, C. Boxe, and E.T. Aaltonee. "Keeping mars warm with new super greenhouse gases." *Proceedings of the National Academy of Sciences of the United States of America* 98, no. 5 (2001), 2154–2157.

Gilder, Carrie. "Space station leaves 'microbial footprint' on astronauts." *NASA.gov*. Last modified 7 May 2020. https://www.nasa.gov/mission_pages/station/research/news/microbiome-space-station-leaves-microbial-fingerprint-on-astronauts.

Golden, Christie. *Star Wars: Dark Disciple*. New York: Del Rey, 2015.

Gopal, Murali, and Alka Gupta. "A sci-fi film about a Mars survivor calls our attention to the importance of microbiome in agriculture." *Current Science* 110, no. 1 (2016), 15–16.

Gormley, Michael J. "The living force: An ecological reading of how the force regards its adherents." *Star Wars and Religion*, special issue of *The Journal of Religion and Popular Culture* 31, no. 1 (2019), 31–43.

Greenwood, Willard P. *Reading Cormac McCarthy*. Santa Barbara: Greenwood Press, 2009.

Hagman, David, Emily H. Ho, and George Lowenstein. "Nudging out support for a carbon tax." *Nature Climate Change* 9 (2019), 484–489.

Haraway, Donna J. *Staying with the Trouble: Making Kin in the Chthulucene*. Durham and London: Duke University Press, 2016.

Hawking, Stephen. *A Brief History of Time*. New York: Bantam Books Trade Paperbacks, 1988.

Hawking, Stephen, and Leonard Mlodinow. "The elusive theory of everything." *Scientific American* 304, no. 4 (2010). Accessed 21 March 2021. https://www.sci entificamerican.com/article/the-elusive-thoery-of-everything/.

Hayles, N. Katherine. *How We Became Posthuman: Virtual Bodies in Cybernetics, Literature, and Informatics*. Chicago: University of Chicago Press, 1999.

Howard, Ron. *Apollo 13*. Universal City, CA: Universal, 1995.

———. *Solo: A Star Wars Story*. San Francisco, CA: Lucasfilm, 2018.

Howell, Steve B., and David Lee Summers, editors. *A Kepler's Dozen: Thirteen Stories About Distant World's that Really Exist*. Mesilla Park, NM: Hadrosaur Productions, 2013.

International Geosphere-Biosphere Programme (IGBP). "Planetary dashboard shows 'Great Acceleration' in human activity since 1950." 15 January 2015. Accessed 19 March 2021. http://www.igbp.net/news/pressreleases/pressreleases/planeta rydashboardshowsgreataccelerationinhumanactivitysince1950.5.950c2fa1495db70 81eb42.html.

International Potato Center. "Indicators show potatoes can grow on Mars." 8 March 2017. Accessed 7 March 2021. https://cipotato.org/blog/indicators-show-potatoes -can-grow-mars/.

Jakobson, Roman. *Language in Literature*. Cambridge: Harvard University Press, 1987.

Kaila, Ville R.I., and Arto Annila. "Natural selection for least action." *Proceedings of the Royal Society A* 464 (July 2008), 3055–3070.

Kaku, Michio. *Physics of the Impossible: A Scientific Exploration into the World of Phasers, Force Fields, Teleportation and Time Travel*. New York: Doubleday, 2008.

———. "Should we use comets to terraform mars?" *Big Think*. 13 September 2010. Accessed 10 March 2021. https://bigthink.com/dr-kakus-universe/should-we-use -comets-and-asteroids-to-terraform-mars.

Kolbert, Elizabeth. *The Sixth Extinction: An Unnatural History*. New York: Picador, 2014.

Lamplugh, Rick. *In the Temple of Wolves: A Winter's Immersion in Yellowstone*. Scotts Valley: CreateSpace Independent Publishing Platform, 2014.

Le Guin, Ursula K. "The ones who walk away from Omelas." In: *Of the Wind's Twelve Quarters*. New York: Harper Perrenial, 1975.

Leopold, Aldo. *A Sand County Almanac*. New York: Oxford University Press, 1949.

Levander, Caroline, and Walter Mignolo. "Introduction: The global south and world dis/order." *The Global South* 5, no. 1 (2011), 1–11.

Liebenberg, Louis. *The Art of Tracking: The Origin of Science*. Claremont, South Africa: David Philip Publishers, 1990.

———. *The Origin of Science: The Evolutionary Roots of Scientific Reasoning and its Implications for Citizen Science*. Capetown, South Africa: CyberTracker, 2013.

Lopez, Barry. *Of Wolves and Men*. New York: Scribner, 1978.

Lucas, George. *Star Wars: The Empire Strikes Back*. Twentieth Century Fox, 1980.

Mach, Katharine, Caroline M. Kraan, W. Neil Adger, Halvard Buhaug, Marshall Burke, James D. Fearon, Christopher B. Field, Cullen S. Hendrix, Jean-Francois Maystadt, John O'Loughlin, Philip Roessler, Jürgen Scheffran, Kenneth A. Schultz, and Nina von Uexkull. "Climate as a risk factor for armed conflict." *Nature* 571 (2019), 193–197.

Malouf, David. *An Imaginary Life*. New York: Random House, 1978.

Mann, Michael E. *The New Climate War: The Fight to Take Back Our Planet*. New York: PublicAffairs, 2021.

Mann, Michael E., R.S. Bradley, and M.K. Hughes. "Global-Scale Temperature Patterns and Climate Forcing Over the Past Six Centuries." *Nature* 392 (1998), 779–787.

MARS. "Novo Mundo." Directed by Everardo Gout. Written by Karen Janszen and Paul Solet. *National Geographic*, 14 November 2016.

Martinez Pastur, Guillermol, M. Vanessa Lencinas, Julio Escobar, Paula Quiroga, Laura Malmierca, and Marta Lizarralde. "Understorey succession in nothofagus forests in tierra del fuego (Argentina) affected by castor canadensis." *Applied Vegetation Science* 9, no. 1 (2006), 143–154.

Marx, Leo. *The Machine in the Garden: Technology and the Pastoral Ideal in America*. New York: Oxford University Press, 1964.

McCarthy, Cormac. *Blood Meridian*. New York: Vintage International, 1985.

———. *The Crossing*. New York: Vintage International, 1994.

McDougall, Christopher. *Born to Run: A Hidden Tribe, Superathletes, and the Greatest Race the World Has Never Seen*. New York: Alfred A. Knopf, 2010.

McDougall, Len. *Tracking and Reading Sign: A Guide to Mastering the Original Forensic Science*. New York: Skyhorse, 2010.

McKibben, Will. *The End of Nature*. New York: Random House, 2006.

McMurtry, Larry. *Streets of Laredo*. New York: Simon and Schuster, 1993.

Miller, Jennifer R.B. "How Trumps wall would alter our biological identity forever: It would destroy an extraordinary web of biodiversity that evolved over millions of years." *Scientific American*. 2 January 2019. Accessed 22 February 2021. https://blogs.scientificamerican.com/observations/how-trumps-wall-would-alter-our-biological-identity-forever/.

Miranda, Lin-Manuel. 2016. "How far I'll go." Track 4 on *Moana: Original Motion Picture Soundtrack*, Walt Disney.

Moore, Allen, and Dave Gibbons. *Watchmen*. New York: DC Comics, 1986.

Moore, Kathleen Dean. *The Pine Island Paradox: Making Connections in a Disconnected World*. Minneapolis, MN: Milkweed Editions, 2005.

Morrison, Grant. *Batman: Arkham Asylum: A Serious House on Serious Earth*. New York: DC Comics, 1989.

Morrow, David, and Toby Svoboda. "Geoengineering and non-ideal theory." *Public Affairs Quarterly* 30, no. 1 (2016), 83–102.

Morton, Timothy. *Hyperobjects: Philosophy and Ecology After the End of the World*. Minneapolis: University of Minnesota Press, 2013.

———. *The Ecological Thought*. Cambridge: Harvard University Press, 2010.

NASA. "Can plants grow with mars soil?" Last modified 7 August 2017a. Accessed 7 March 2021. https://www.nasa.gov/feature/can-plants-grow-with-mars-soil.

———. "Life support systems." Last modified 3 August 2017b. Accessed 28 February 2021. https://www.nasa.gov/content/life-support-systems.

———. "NASA spacecraft images offer sharper views of apollo landing sites." Last modified 7 August 2017c. Accessed 28 February 2021. https://www.nasa.gov/miss ion_pages/LRO/news/apollo-sites.html.

———. "National space exploration campaign report." September 2018.

———. "Deep impact (EPOXI)." Last modified 24 July 2019a. Accessed 10 March 2021. https://solarsystem.nasa.gov/missions/deep-impact-epoxi/in-depth/.

———. "Near shoemaker." Last modified 31 July 2019b. Accessed 10 March 2021. https://solarsystem.nasa.gov/missions/near-shoemaker/in-depth/.

———. "6 Technologies NASA is advancing to send humans to mars." Last modified 20 July 2020a. Accessed 7 March 2021. https://www.nasa.gov/directorates/spacet ech/6_Technologies_NASA_is_Advancing_to_Send_Humans_to_Mars.

———. "Artemis plan: NASA's lunar exploration program overview." September 2020b.

———. "Moon to mars." Last modified 25 September 2020c. Accessed 27 February 2021. https://www.nasa.gov/topics/moon-to-mars.

———. "Solar power investigation to launch on lunar lander." Last modified 16 July 2020d. https://www.nasa.gov/feature/glenn/2020/solar-power-investigation-to -launch-on-lunar-lander.

———. "The human body in space." Last modified 5 February 2021. Accessed 28 February 2021. https://www.nasa.gov/hrp/bodyinspace.

NBC Montana. "Wolves in Yellowstone National Park." *Facebook*. 6 January 2020. Accessed 13 March 2021. https://www.facebook.com/watch/?v=52958103764 5891&extid=12Wj54UG2RhwSbAE.

Nelson, Melissa K. "Conclusion: Back in our tracks – Embodying kinship as if the future mattered." In *Traditional Ecological Knowledge: Learning from Indigenous Practices for Environmental Stability*, edited by Melissa K. Nelson and Dan Shilling. Cambridge: Cambridge University Press, 2018, 250–262.

Newcomb, Simon. "Modern mathematical thought." *Bulletin of the New York Mathematical Society* 3 (1894), 95–107.

Nixon, Rob. *Slow Violence and the Environmentalism of the Poor*. Cambridge, MA: Harvard University Press, 2011.

Nolan, Christopher. *Inception*. Burbank, CA: Warner Home Video, 2010.

————. *Interstellar*. Burbank, CA: Warner Home Video, 2014.

————. *The Dark Knight*. Burbank, CA: Warner Home Video, 2008.

Núñez, Rafael E., and Eve Sweetser. "With the future behind them: Convergent evidence from aymara language and gesture in the crosslinguistic comparison of spatial construals of time." *Cognitive Science* 30 (2006), 401–450.

Oberhaus, Daniel. "A crashed israeli lunar lander spilled tardigrades on the moon." *Wired*. Last modified 6 August 2019. Accessed 14 March 2021. https://www.wired.com/story/a-crashed-israeli-lunar-lander-spilled-tardigrades-on-the-moon/.

"Old Ephraim." Utah State University Digital History Collections. Accessed 27 January 2021. https://digital.lib.usu.edu/digital/collection/Ephraim.

Oxford Dictionary. "Replicate." Accessed 27 August 2020. https://www.lexico.com/en/definition/replicate.

Oxford English Dictionary. "Replicate." Accessed 27 August 2020. https://www-oed-com.ezproxy.bpl.org/view/Entry/162879?rskey=mu7C4n&result=1#eid.

Oxford University Press. "The OED and Oxford dictionaries." Accessed 27 August 2020. https://www.oed.com/page/oedodo/The+OED+and+Oxford+Dictionaries.

Pak, Chris. *Terraforming: Ecopolitical Transformations and Environmentalism in Science Fiction*. Liverpool: Liverpool University Press, 2016.

Parini, Jay. "The greening of the humanities." *The New York Times*. 29 October 1995. http://www.nytimes.com/1995/10/29/magazine/the-greening-of-the-humanities.html?pagewanted=all.

Peterson, Brenda. *Wolf Nation: The Life, Death, and Return of Wild American Wolves*. Boston: Da Capo Press, 2017.

Petranek, Stephen L. *How We'll Live on Mars*. New York: TED Books, 2015.

Polynesian Voyaging Society. "PVS mission and vision." Accessed 1 March 2021. http://www.hokulea.com/vision-mission/.

Postgate, John. *Microbes and Man*. Cambridge: Cambridge University Press, 2000.

Powici, Chris. "Witnessing the wolf the human and the lupine in Cormac McCarthy's *The Crossing*." *Postgraduate English* 1 (March 2000), 1–13.

"Rattlestar Ricklactica." *Rick and Morty*, Directed by Wesley Archer and Jacob Hair, Written by Justin Roiland and Dan Harmon, season 4, episode 5, 15 December. Williams Street.

Read, Carveth. *The Origin of Man*. Cambridge: Cambridge University Press, 1920.

Reed, Robert. "A place with shade." *Worldmakers: SF Adventures in Terraforming*. New York: St. Martin's Griffin, 2001, 193–220.

Rezendes, Paul. *Tracking and the Art of Seeing: How to Read Animal Tracks and Sign*. New York: HarperCollins, 1999.

Ridley, Scott. *The Martian*. Los Angeles, CA: Twentieth Century Fox, 2015.

Ripple, William J., and Robert L. Beschta. "Restoring Yellowstone's aspen with wolves." *Biological Conservation* 138 (2007), 514–519. doi: 10.1016/j.biocon.2007.05.006.

Rivers, Nathaniel A. "Better footprints." In: *Tracing Rhetoric and Material Life*. Camden: Palgrave Macmillan, 2018, 169–196.

Romero, James. "Turning the Red Planet green: How we'll grow crops on Mars." *BBC Science Focus*. 17 October 2020. Accessed 7 March 2021. https://www.sci

encefocus.com/space/turning-the-red-planet-green-how-well-grow-crops-on-mar s/.

Rueckert, William. "Literature and ecology: An experiment in ecocriticism." Rpt. from 1978 in *The Ecocriticism Reader: Landmarks in Literary Ecology*, edited by Cheryll Glotfelty and Harold Fromm. Athens: University of Georgia Press, 1996, 105–123.

Russo, Joe, and Anthony Russo. *Avengers: Endgame*. Burbank: Disney, 2019.

———. *Avengers: Infinity War*. Burbank: Disney, 2018.

Sagan, Carl. *Contact*. New York: Pocket Books, 1985.

Sagan, Dorion, and Lynn Margulis. *Garden of Microbial Delights: A Practical Guide to the Subvisible World*. Dubuque: Kendall Hunt, 1993.

Schaberg, Christopher. *Searching for the Anthropocene: A Journey into the Environmental Humanities*. New York: Bloomsbury Academic, 2020.

Scharf, Caleb A. "Tardigrades were already on the moon." *Scientific American: Life, Unbounded*. 8 August 2019. Accessed 14 March 2021. https://blogs.scientifica merican.com/life-unbounded/tardigrades-were-already-on-the-moon/.

Schreiber, Alexander, and Steven Gimbel. "Evolution and the second law of thermo-dynamics: Effectively communicating to non-technicians." *Evolution: Education and Outreach* 3 (January 2010), 99–106.

Schwaegerl, Christian. "The anthropocene: Paul Crutzen's epochal legacy." *Anthropocene Magazine*, translated Margaret Ries. Published 14 February 2021. Accessed 2 March 2021. https://www.anthropocenemagazine.org/2021/02/the-a nthropocene-paul-crutzens-epochal-legacy/.

Schwartz, James S.J. "On the moral permissibility of terraforming." *Ethics and the Environment* 18, no. 2 (2013), 1–31.

Seife, Charles. *Sun in a Bottle: The Strange History of Fusion and the Science of Wishful Thinking*. New York: Viking, 2008.

Shamsian, Jacob. "Fact-checking 'The Martian': Can you really grow plants on mars?" *Modern Farmer*. 2 October 2015. Accessed 7 March 2021. https://modernf armer.com/2015/10/can-you-grow-plants-on-mars/.

Shelley, Mary. *Frankenstein; or The Modern Prometheus (1818)*. Edited by D.L. Macdonald and Kathleen Scherf. Claremont, Canada: Broadview Press, 2012.

Smith, Philip. "The American Yeoman in Andy Weir's *The Martian*." *Science Fiction Studies* 46, no. 2 (2019), 322–341.

Song, Se Jin, et al. "Cohabiting family members share microbiota with one another and with their dogs." *eLife* 2, no. e00458 (April 2013). doi: 10.7554/eLife.00458.

Spudis, Paul D. "Faded flags on the moon: The probable current state of the Apollo American flags on the Moon: Symbolic?" *Air & Space Magazine*. Published 19 July 2011. Accessed 28 February 2021. https://www.airspacemag.com/daily-planet /faded-flags-on-the-moon-32929921/.

Stanton, Andrew. *Wall-E*. Burbank, CA: Walt Disney Home Entertainment, 2008.

Steffen, Will, Paul J. Crutzen, and John R. McNeill. "The anthropocene: Are humans now overwhelming the great forces of nature?" *Ambio* 36, no. 8 (2007), 614–621.

Steffen, Will, Wendy Broadgate, Lisa Deutsch, Owen Gaffney, and Cornelia Ludwig. "The trajectory of the Anthropocene: The great acceleration." *The Anthropocene Review* 2, no. 1 (2015), 81–98.

Stephenson, Neal. *Seveneves*. New York: William Morrow, 2015.

Tang, Mimi L.K., and Raymond J. Mullins. "Food allergy: Is prevalence increasing?" *Internal Medicine* 47, no. 3 (2017), 256–261.

"The Focus Attenuation." In: *The Big Bang Theory*, directed by Mark Cendrowski, written by Chuck Lorre and Bill Prady, season 8, episode 5, 13 October. Chuck Lorre Productions, 2015.

The Planetary Society. "Flight by light for CubeSats." Accessed 7 March 2021. https://www.planetary.org/sci-tech/lightsail.

Tolkien, J.R.R. *The Lord of the Rings: The Two Towers*. New York: Ballantine Books, 1954.

Trivedi, Bijal. "The wipeout gene." *Scientific American* (November 2011), 68–75.

United States Space Force. "Mission." Accessed 13 March 2021. https://www.spaceforce.mil/About-Us/About-Space-Force/Mission/.

Urry, John. *Consuming Places*. New York: Routledge, 1995.

Urry, John, and Jonas Larsen. *The Tourist Gaze 3.0*. Los Angeles: SAGE, 2011.

U.S. Commercial Space Launch Competitiveness Act. Public Law 114-90. *US Statutes at Large* (2015).

VanderMeer, Jeff. *Acceptance*. New York: Farrar, Strauss and Giroux, 2014a.

———. *Annihilation*. New York: Farrar, Strauss and Giroux, 2014b.

———. *Authority*. New York: Farrar, Strauss and Giroux, 2014c.

Villeneuve, Denis. *Arrival*. Los Angeles, CA: Paramount, 2015.

Vonnegut, Kurt. *Slaughterhouse-Five*. New York: Dial Press, 1969.

Voorhies, Alexander, et al. "Study of the impact of long-duration space missions at the International Space Station on the astronaut microbiome." *Scientific Reports* 9, no. 9911 (July 2019). doi: 10.1038/s41598-019-46303-8.

Wageningen Environmental Research. "Earthworms can reproduce in Mars soil simulant." *Wageningen University and Research*. 27 November 2017. Accessed 7 March 2021. https://www.wur.nl/en/newsarticle/Earthworms-can-reproduce-in-Mars-soil-simulant.htm.

———. "First bean harvest on martian and lunar soil fertilised with urine." *Wageningen University and Research*. 21 January 2021. Accessed 7 March 2021. https://www.wur.nl/en/Research-Results/Research-Institutes/Environmental-Research/show-wenr/First-bean-harvest-on-martian-and-lunar-soil-fertilised-with-urine.htm.

Wamelink, G.W.W., Joep Y. Frissel, Wilfred H.J. Krijnen, M. Rinie Verwoert, and Paul W. Goedhart. "Can plants grow on mars and the moon: A growth experiment on mars and moon soil simulants." *PLoS One* 9, no. 8 (2014).

Waters, Colin N., et al. "Can nuclear weapons fallout mark the beginning of the Anthropocene Epoch?" *Bulletin of the Atomic Scientists* 71, no. 3 (2015), 46–57.

Weir, Andy. *The Martian*. New York: Broadway Books, 2014.

Wells, H.G. *The Time Machine*. Edited by Nicholas Ruddick. Ontario: Broadview Press, 2001.

———. *War of the Worlds*. New York: Barnes and Noble Classics, 2004.

White, T.H. *The Once and Future King*. New York: Penguin, 1939.

Wicken, Jeffrey S. "Entropy, information, and nonequilibrium evolution." *Systematic Zoology* 32, no. 4 (1983), 438–443.

Williams, Raymond. *The Country and the City*. New York: Oxford University Press, 1973.

Wilson, Edward O. *Consilience: The Unity of Knowledge*. New York: Vintage Books, 1998.

xkcd. "The Martian." Accessed 21 May 2020. https://xkcd.com/1536.

"Yellowstone National Park." National Parks Foundation. Accessed 27 February 2021. https://www.nationalparks.org/explore-parks/yellowstone-national-park.

Young, Vincent B. "The role of the microbiome in human health and disease." *British Medical Journal* 356 (2017), 1–14.

Zeh, H.D. *The Physical Basis of the Direction of Time*. Germany: Springer, 2001.

Index

About the Author

Courtesy of Abigail West.

Michael J. Gormley is an ecocritic and professor at Quinsigamond Community College in Worcester, Massachusetts. His research and teaching focus on tracking in literature, nonlinear temporality, the intersections of mathematics and science with the humanities, and nerdy pop culture stuff—especially *Star Wars*. His article, "The Living Force: An Ecological Reading of How the Force Regards Its Adherents" brings most of that together. For roughly ten years now, Gorm has trained in wilderness survival rooted in animal tracking, which he developed into an ecocritical framework for literary analysis and his broader perspective in the environmental humanities.

Lightning Source UK Ltd.
Milton Keynes UK
UKHW010225210721
387494UK00001B/15